MEDIA AND COMMUNICATIONS - TECHNOLOGIES, POLICIES AND CHALLENGES

SATELLITE COMMUNICATIONS

SERVICES INDUSTRY COMPETITION AND COMMERCIAL RELIANCE ON THE GPS

MEDIA AND COMMUNICATIONS - TECHNOLOGIES, POLICIES AND CHALLENGES

Additional books in this series can be found on Nova's website under the Series tab.

Additional E-books in this series can be found on Nova's website under the E-book tab.

MEDIA AND COMMUNICATIONS - TECHNOLOGIES, POLICIES AND CHALLENGES

SATELLITE COMMUNICATIONS

SERVICES INDUSTRY COMPETITION AND COMMERCIAL RELIANCE ON THE GPS

Bret G. Osborn
and
Emil T. Rowland
EDITORS

Nova Science Publishers, Inc.
New York

Library of Congress Cataloging-in-Publication Data

ISBN 978-1-61942-464-7

Published by Nova Science Publishers, Inc. † New York

CONTENTS

PREFACE

Commercial satellites are used to provide a variety of fixed satellite services, ranging from consumer satellite television and broadband to military communications in remote regions. This book examines the changes that have occurred in the fixed satellite services industry since 2000 and the effects these changes could have on the relationship between satellite operators and service providers. Also discussed is the commercial communications industry's use of and reliance on the GPS, in accordance with White House direction to tailor the effort to enable a quick response. The National Security Telecommunications Advisory Committee (NSTAC) notes that reliance on GPS signals in military, maritime, aviation, and other civil environments varies widely depending on the specific use and application.

Chapter 1- Commercial satellites are used by the U.S. government to provide a variety of fixed satellite services, such as military communications. However, the number of satellite operators providing such service has declined since 2000. Further, until recently, three vendors, known as satellite service providers, had sole authority to contract with the Department of Defense (DOD) under its primary satellite contract.

Among other things, GAO was asked to describe (1) changes that have occurred in the fixed satellite services industry since 2000 and the effects these changes could have on the relationship between satellite operators and service providers; (2) the technological, regulatory, and other factors that affect competition in the fixed satellite services industry; and (3) how costs for DOD to acquire fixed satellite services have changed since 2000 and contracting officials' views on the effects of changes in the industry and contracts on costs. To address these objectives, GAO reviewed demand and capacity data, economic literature, and international regulations; analyzed DOD commercial satellite procurement data; and interviewed satellite industry stakeholders.

The Federal Communications Commission, DOD, General Services Administration, and the Departments of Justice and State reviewed a draft of this report. The agencies provided technical comments that GAO incorporated into the report as appropriate.

Chapter 2- The President's National Security Telecommunications Advisory Committee (NSTAC) performed an evaluation in response to a White House request for commercial communications industry findings on the commercial communications infrastructure's reliance on the Global Positioning System (GPS). To gain current perspectives on the industry-wide use of GPS, the NSTAC solicited information from its members, other providers within the industry, and several external subject matter experts. Specifically, the NSTAC requested information on: (1) company and industry segment use of and reliance on GPS signals; (2) impacts to networks and operations that would result from loss or degradation of GPS signals; and (3) specific strategies implemented or planned to mitigate the impact of any GPS signal loss or degradation.

In: Satellite Communications ISBN: 978-1-61942-464-7
Editors: B. G. Osborn & E. T. Rowland © 2012 Nova Science Publishers, Inc

Chapter 1

TELECOMMUNICATIONS: COMPETITION, CAPACITY, AND COSTS IN THE FIXED SATELLITE SERVICES INDUSTRY*

United States Government Accountability Office

WHY GAO DID THIS STUDY

Commercial satellites are used by the U.S. government to provide a variety of fixed satellite services, such as military communications. However, the number of satellite operators providing such service has declined since 2000. Further, until recently, three vendors, known as satellite service providers, had sole authority to contract with the Department of Defense (DOD) under its primary satellite contract.

Among other things, GAO was asked to describe (1) changes that have occurred in the fixed satellite services industry since 2000 and the effects these changes could have on the relationship between satellite operators and service providers; (2) the technological, regulatory, and other factors that affect competition in the fixed satellite services industry; and (3) how costs for DOD to acquire fixed satellite services have changed since 2000 and contracting

* This is an edited, reformatted and augmented version of the United States Government Accountability Office publication, Report to Congressional Requesters, GAO-11-777, dated September 2011.

officials' views on the effects of changes in the industry and contracts on costs. To address these objectives, GAO reviewed demand and capacity data, economic literature, and international regulations; analyzed DOD commercial satellite procurement data; and interviewed satellite industry stakeholders.

The Federal Communications Commission, DOD, General Services Administration, and the Departments of Justice and State reviewed a draft of this report. The agencies provided technical comments that GAO incorporated into the report as appropriate.

What GAO Found

Since 2000, integration in the fixed satellite services industry has occurred, altering the relationships between satellite operators (owners of satellites) and service providers (resellers of satellite services). Mergers within the industry have resulted in two primary operators—Intelsat and SES—providing service to the United States. In addition, these two satellite operators also acquired U.S. subsidiaries, allowing them to compete against the service providers for government contracts. According to service providers, this change could result in an operator charging a higher price for capacity to the service provider than to its subsidiary, placing the service provider at a competitive disadvantage. Alternatively, the operator may be able to provide capacity more efficiently and at a lower cost than if the customer, such as the U.S. government, acquired the capacity indirectly through a service provider.

A limited number of orbital locations and enforcement mechanisms in international regulations constrain entry into the fixed satellite services industry. A finite number of orbital locations limit the number of satellites in orbit. International regulations have processes in place to promote equitable and efficient access to orbital resources, but the International Telecommunication Union, a United Nations specialized agency, does not have the ability to monitor and enforce these regulations. As a result, administrations (countries) file numerous applications for orbital locations that may not result in the launch of a satellite, preventing other operators from entering the industry due to limited overall slots.

DOD's costs to acquire fixed satellite services have increased significantly since 2003, but contracting officials expect a new contract to increase competition. According to GAO's analysis of DOD data, the real cost per megahertz of bandwidth was 30 percent lower in fiscal year 2003, and lower in all intervening years, than in fiscal year 2010. Contracting officials attribute the higher costs to market factors, such as demand and availability of bandwidth, and expect a new government contract to increase competition, which may exert downward pressure on the government's costs.

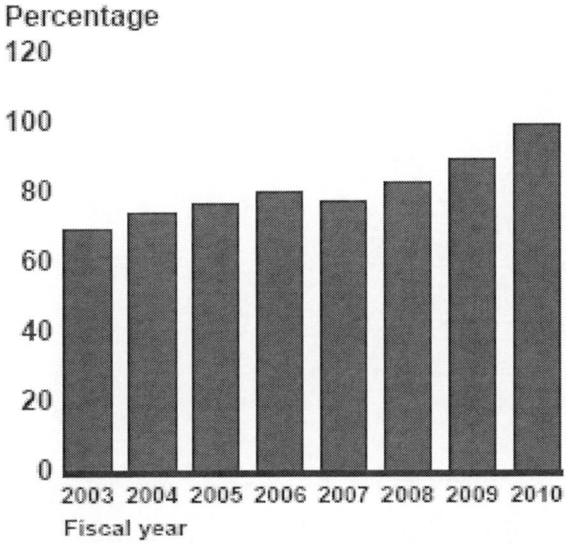

Source: GAO analysis of DOD data.

Note: GAO derived these estimates from a regression model using data from 2001 through March 2011. Data for 2001, 2002, and 2011 were omitted from this figure as the estimates were not statistically significant at the 5 percent confidence level.

Estimated Real Cost per Megahertz of Bandwidth for DOD Task Orders as a Percentage of Estimated Costs for Fiscal Year 2010.

ABBREVIATIONS

DISA	Defense Information Systems Agency
DOD	Department of Defense
DOJ	Department of Justice
DSTS-G	Defense Information Systems Network Satellite Transmission Services-Global
FCC	Federal Communications Commission
FCSA	Future Commercial Satellite Communications Services Acquisition program
GSA	General Services Administration

HDTV	high-definition television
ITU	International Telecommunication Union
MHz	megahertz
ORBIT	Open-market Reorganization for the Betterment of International Telecommunications
SATCOM-II	Satellite Communications II
TPE	transponder equivalents
USSTRATCOM	United States Strategic Command

September 7, 2011
The Honorable Mark Pryor
The Honorable John Thune
The Honorable David Vitter
The Honorable Mark Warner
United States Senate
The Honorable Gerry Connolly
House of Representatives

Commercial satellites are used to provide a variety of fixed satellite services, ranging from consumer satellite television and broadband to military communications in remote regions.[1] According to the Satellite Industry Association, revenues in the satellite services sector, which includes fixed satellite services, increased 9 percent from 2009 to 2010, from $93 billion to $101.3 billion.[2] Revenues from fixed satellite services increased from $14.4 billion to $15 billion from 2009 to 2010. The majority of these revenues are from transponder agreements, in which customers lease bandwidth capacity directly or indirectly from a satellite operator.[3] Commercial users, such as broadcast networks, are the largest consumers of fixed satellite bandwidth. However, the U.S. government also uses commercial satellites for military operations, satellite imagery, distance learning programs, and disaster recovery response efforts, among other things.

Within the federal government, the Department of Defense (DOD) is the largest user of commercial satellite bandwidth. The department is increasingly relying on commercial satellites to meet demand unable to be met by its military communications satellites, particularly in support of ongoing operations in the Middle East. According to DOD, its use of commercial fixed satellite bandwidth in the Middle East and Africa increased by more than 180 percent from 2003 through 2009, and the region accounted for over half of

DOD's overall bandwidth usage in fiscal year 2009.[4] At the same time, bandwidth expenditures in the region increased by more than 170 percent. DOD expects its use of commercial satellite bandwidth to increase, particularly to support an increased demand for airborne intelligence, surveillance, and reconnaissance assets.[5]

As DOD relies more on commercial satellites to support its operations, the availability and affordability of commercial satellite bandwidth will be important in allowing DOD to continue to meet its global communications needs. Until recently, three vendors, known as satellite service providers, held the primary contract DOD used to meet its commercial bandwidth needs; the Defense Information Systems Network Satellite Transmission Services-Global (DSTS-G) contract was the primary contract used by DOD to acquire fixed satellite services and was a total set-aside for competition restricted to small business concerns. Now, satellite operators, which own the satellites, are also eligible to supply bandwidth directly to the department. However, the number of satellite operators providing service globally and to the United States has declined since 2000 because of mergers and acquisitions. Today, there are four major satellite operators, two of which primarily provide bandwidth capacity to the United States. With fewer operators providing global coverage, concerns have arisen about the availability and affordability of commercial bandwidth capacity for the government to continue to meet its global communications needs.

Congress passed the Open-market Reorganization for the Betterment of International Telecommunications (ORBIT) Act in 2000 to promote a more competitive global satellite services market.[6] Because the government's reliance on commercial satellites has increased since the ORBIT Act was passed, you asked us to assess the state of competition in the fixed satellite services industry. Accordingly, this report describes (1) how current satellite capacity compares to current and forecasted demand; (2) the changes that have occurred in the fixed satellite services industry since 2000 and the effects these changes could have on the relationship between satellite operators and service providers; (3) technological, regulatory, and other factors that affect competition in the fixed satellite services industry; and (4) how costs for DOD to acquire fixed satellite services have changed since 2000 and contracting officials' views on the effects of changes in the industry and contracts on costs.

To address these objectives, we reviewed and analyzed Federal Communications Commission (FCC) reports on the ORBIT Act and competition in the satellite industry. We also interviewed satellite industry stakeholders, including satellite operators and service providers, as well as government officials from DOD, FCC, the General Services Administration (GSA), the Department of Justice (DOJ), and the Department of State. To determine how current satellite capacity compares to current and forecasted demand for fixed satellite services, we used data from Futron Corporation, a research firm with expertise in the satellite industry, on forecasted satellite demand; we examined the procedures and controls that Futron used to gather and report its data and determined that the data were sufficiently reliable for the purposes of our report. We also analyzed satellite operators' annual financial reports and interviewed financial analysts that track the satellite industry. To determine the changes that have occurred in the industry since 2000 and the effects these changes could have on the relationship between satellite operators and service providers, we reviewed comments filed for FCC reports, analyzed economic literature on competition, and interviewed economists with expertise in horizontal and vertical integration, including officials from DOJ's Antitrust Division. To determine the technological, regulatory, and other factors that affect competition in the fixed satellite services industry, we reviewed International Telecommunication Union (ITU) regulations and interviewed ITU and State Department officials as well as industry officials with regulatory expertise. To determine how DOD's costs to acquire fixed satellite services have changed since 2000 and contracting officials' views on the effects of changes in the industry and contracts on costs, we analyzed DOD data on bandwidth costs under DOD's primary contract from 2001 through March 2011 and reviewed DOD reports on the department's usage of and expenditures for commercial satellite communications services; we also examined DOD's processes to ensure the reliability of its data and determined that the data were sufficiently reliable for our report. Finally, we interviewed officials from DOD and GSA regarding the effects of changes in the industry and contracts on the cost to acquire satellite service.

We conducted our work from September 2010 to September 2011 in accordance with all sections of GAO's Quality Assurance Framework that are relevant to our objectives. The framework requires that we plan and perform the engagement to obtain sufficient and appropriate evidence to meet our stated objectives and to discuss any limitations in our work. We believe that the information and data obtained, and the analysis conducted, provide a reasonable basis for any findings and conclusions.

BACKGROUND

Most fixed satellite service satellites operate in geostationary orbit more than 22,000 miles above the equator.[7] From this position, a satellite appears stationary over a location on the earth and can theoretically provide coverage to about one-third of the earth's surface. Fixed satellite service satellites communicate with ground infrastructure by receiving signals from earth stations or antennas and retransmitting the signals to other locations on the earth's surface through transponders on the satellite.[8] A typical satellite has 24 to 72 transponders. Satellites can transmit signals in a variety of configurations. A common configuration supported by fixed satellite service satellites is a point-to-multipoint configuration, in which a signal from a single earth station is sent to a satellite and then retransmitted from the satellite to multiple sites. For example, a signal from a broadcast network is transmitted to the satellite and then retransmitted to individual broadcast stations. Figure 1 illustrates the basic functions of a satellite operating in geostationary orbit.

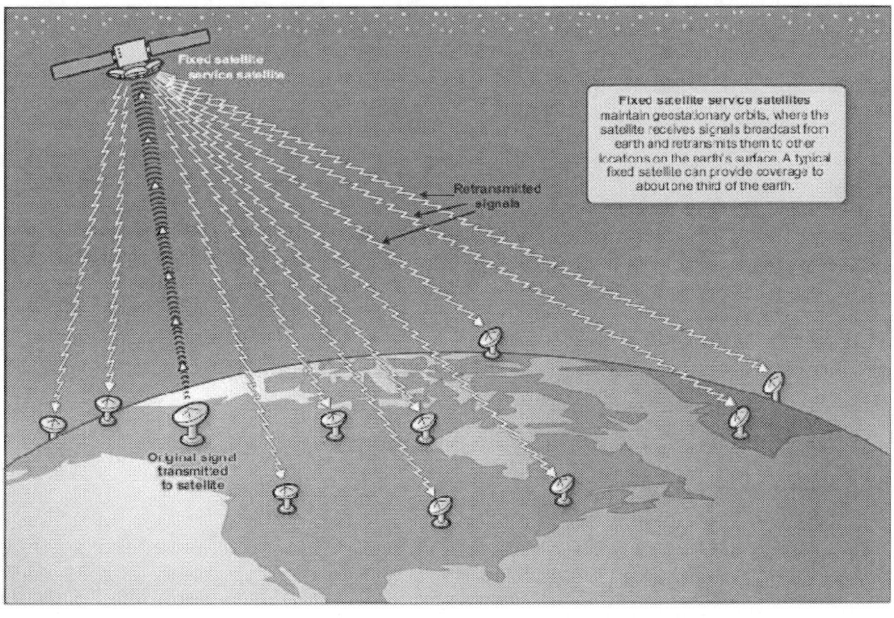

Source: GAO.

Figure 1. Illustration of a Point-to-Multipoint Fixed Satellite Transmission Operating in Geostationary Orbit.

The transponders on a fixed satellite service satellite transmit radio signals to the earth using radio spectrum, in assigned frequency bands. Given their stationary position and ability to provide coverage to a large geographic area, fixed satellite service satellites are commonly used to provide a variety of communications services. Four frequency bands—C, Ku, Ka, and X—are most commonly associated with fixed satellite services. Each of these frequency bands has certain advantages and disadvantages for various applications. For example, transponders operating at lower C-band frequencies are useful to broadcast television networks distributing video content to local broadcast stations; C-band frequencies are less susceptible to degradation from precipitation than other bands.[9] By contrast, Ku-band transponders operate at higher frequencies than C-band transponders and can therefore communicate with smaller dishes and offer more flexibility for customers.[10] The military primarily uses Ku-band satellites because the dishes offer more mobility than C-band satellite dishes. The Ku-band is also used for satellite news gathering, television network distribution, and corporate enterprise networks, including those used for point-of-sale retail transactions, through what are known as Very Small Aperture Terminals, or VSAT, networks.[11] Portions of this band are used exclusively for satellite television to provide service through designated broadcast satellites.[12] The still-higher-frequency Ka-band has recently begun to play a role in the fixed satellite services industry. Ka-band satellites can transmit more data than C- and Ku-band satellites and are capable of having a large number of beams focused on the earth's surface.[13] However, while Ka-band satellites can provide services to smaller dishes, their signals are more susceptible to degradation from rain than satellites that use lower frequency bands. Currently being used to provide consumer broadband service and television to the home, Ka-band satellites are also planned for use by the maritime and oil and gas industries. In the United States, the X-band is specifically designated for use by the U.S. government and the North Atlantic Treaty Organization. Figure 2 highlights some of the common and emerging uses of fixed satellite services bandwidth.

Geostationary satellites are operated in particular orbital locations in the geostationary arc. The use of the geostationary arc is coordinated by ITU members using a process overseen by ITU. That is, for the location of a satellite to obtain international recognition, its orbital location and frequency assignment must be registered with ITU. ITU is the United Nations specialized agency for information and communication technologies and has 192 member administrations (countries). ITU was founded to promote international cooperation and develop consensus positions within the telecommunications and information technology industries. The Radiocommunication Bureau within ITU manages the coordination and recording procedures for satellite systems and earth stations by publishing data and recording frequency assignments in the Master International Frequency Register (master register). According to ITU, its Radio Regulations is a treaty that provides for the rational, efficient, and equitable use of orbital locations and frequencies as agreed to by the member administrations.

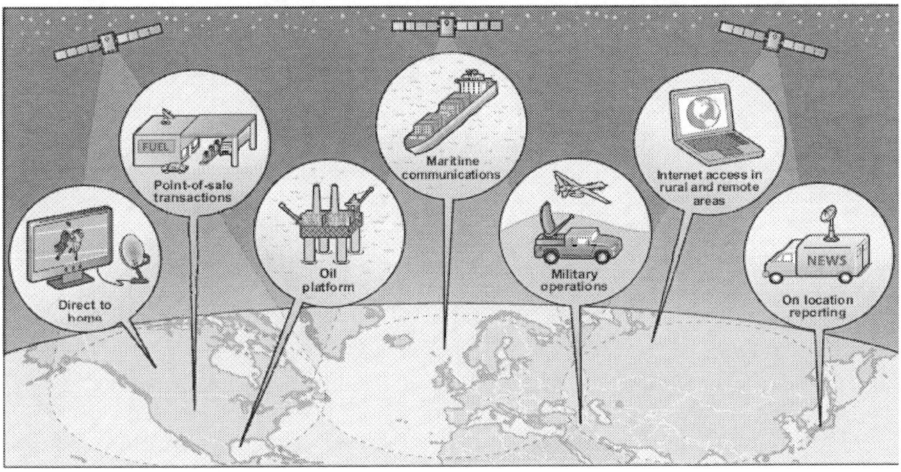

Sources: GAO and Map Resources (Map).

Figure 2. Common and Emerging Uses of Fixed Satellite Services Bandwidth.

Once a satellite has been launched into orbit, users of fixed satellite services acquire bandwidth capacity either directly from a satellite operator or indirectly through a satellite service provider. Satellite operators own and operate satellites and lease bandwidth capacity to customers on a wholesale basis. Satellite operators also occasionally provide other services, such as access to teleports, which connect terrestrial networks with satellite transponders in orbit, or other infrastructure, to customers, as requested. Other users may acquire needed bandwidth capacity through a third-party satellite service provider. For example, a customer may obtain its capacity through a reseller, which purchases the capacity from the satellite operator and then resells the capacity to the end customer. Alternatively, integrators purchase satellite capacity from satellite operators and then resell this capacity to the end customer along with other value-added services. Figure 3 illustrates the ways in which customers typically acquire fixed satellite services bandwidth capacity.

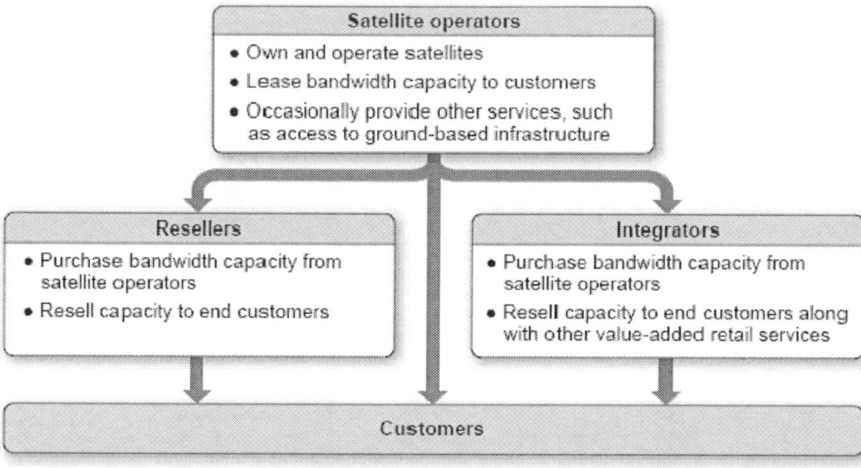

Source: GAO.

Figure 3. Customer Options for Acquiring Fixed Satellite Services Bandwidth.

The federal government acquires satellite capacity through various contracts. The two primary contracts are the Satellite Communications II (SATCOM-II) contract administered by GSA and the DSTS-G contract administered by the Defense Information Systems Agency (DISA) within DOD. In order to provide a common marketplace for the government to procure fixed satellite services, DISA and GSA recently formed a partnership to develop new contracting vehicles through the Future Commercial Satellite Communications Services Acquisition (FCSA) program. The FCSA program consists of a comprehensive set of acquisition activities intended to replace the existing GSA and DISA contracts[14] and will allow the government to procure satellite services in three categories—transponded capacity, subscription services, and endto-end solutions.[15] Vendors may be continuously added to the list of qualified contractors that are able to compete to provide services in the transponded capacity and subscription service categories. As of July 1, 2011, 10 vendors had been approved to compete in the transponded capacity category and 12 vendors had been approved to compete in the subscription services category. GSA and DOD expect the contract for end-to-end solutions to be available in the first quarter of fiscal year 2012.

Congress passed the ORBIT Act in 2000 to promote a more competitive global satellite services marketplace. Specifically, the ORBIT Act provided a U.S. regulatory framework for the privatization of INTELSAT, an intergovernmental organization that provided satellite services. INTELSAT transferred its assets and operating responsibilities to Intelsat, a private company.[16] At the time of its privatization, commercial satellite companies were concerned that INTELSAT enjoyed certain advantages stemming from its previous intergovernmental status that limited the competitiveness of the global satellite marketplace; for example, these companies cited immunity from legal liability as an advantage INTELSAT enjoyed.[17] FCC is required to report annually to Congress on the status of privatization[18] as well as the state of competition in the fixed satellite services industry.[19]

GLOBAL SATELLITE CAPACITY IS SUFFICIENT TO MEET CURRENT AND FORECASTED DEMAND

Current Capacity and Demand

Current fixed satellite capacity is sufficient to meet existing demand on a global basis. According to estimates by Futron, global demand for fixed satellite services is expected to be about 79 percent of available capacity in 2011.[20] Similarly, based on our review of the financial reports for the four major satellite operators, about 82 percent of the capacity on their satellite fleets is currently being used.[21] Three of the satellite operators we interviewed told us that it is important to maximize the utilization of their satellite fleets to maintain a reasonable return on investment. However, satellite operators also cited other factors that are taken into consideration, for example, the desire to have some reserve capacity available if a satellite malfunction occurs or having capacity available to accommodate any increases in fixed satellite services demand from existing or new customers.

Although, on a global basis, fixed satellite capacity is sufficient to meet existing demand, satellite capacity has not always been sufficient to meet DOD's demand in the Middle East, particularly in Iraq and Afghanistan where troops operating in remote regions rely on satellite communications. DOD officials told us that some of their bandwidth requirements have not been met because of capacity constraints, and satellite service providers noted a lack of available satellites in the region that are able to meet DOD's requirements.

FCC, in its most recent report on competition in the fixed satellite services industry, analyzed data from 2002 through 2007 for the Middle East and Africa combined, and reported that the two regions together had less available capacity than other regions, such as North and South America, Asia, and Western Europe.[22] DOD officials said that this regional supply constraint is further exacerbated by the fact that the Department does not lease capacity from all possible global operators, given geopolitical concerns.

Future Capacity and Demand

A variety of factors are contributing to increasing demand for fixed satellite services. We identified four factors that are expected to increase demand and two that are expected to decrease demand for fixed satellite services, with demand increasing overall. The factors expected to increase demand now and in the future include:

- **Increasing use of satellites for broadband Internet access**. Industry officials we spoke with stressed that consumers are increasingly demanding high-speed (broadband) Internet access that can best be provided in remote areas by satellite. Satellite operators are launching new satellites to meet this demand. For example, one major satellite operator launched a new satellite in December 2010, and a satellite product and services company plans to launch a new satellite later in 2011 offering high-speed satellite Internet service.
- **Growth in corporate enterprise networks**. All four of the major satellite operators we spoke with provide satellite services for corporate enterprise networks, and their financial reports indicate continued growth in this market. For example, one major satellite operator reported a 10 percent increase in the number of sites using its VSAT services in just 1 year, due in part to demand from developing markets as well as from government customers.
- **Increase in satellite television and high-definition television (HDTV) channels**. Industry officials that we spoke with told us that demand for HDTV via satellite is growing in developed and emerging markets; HDTV requires greater bandwidth than standard definition format television. A report by the Satellite Industry Association found that the number of HDTV channels worldwide nearly tripled from May 2008 through May 2011 (from 1,353 channels to 3,853

channels).[23] In addition, incremental demand for standard definition format television is expected to continue growing in some emerging markets.

- **Increase in satellite support for military operations.** DOD expects increased demand for satellite capacity to support airborne intelligence, surveillance, and reconnaissance assets. Government and satellite industry officials told us that satellite communication links to ground troops and unmanned aerial vehicles for military operations in Iraq and Afghanistan have increased demand for satellite services in the Middle East. DOD recently requested an increase in the number of combat air patrols for the Air Force's Predator and Reaper unmanned aerial vehicle programs to 50 by fiscal year 2011, an increase of nearly 300 percent since fiscal year 2007.

While the preceding factors increase demand, the following factors can lead to a decrease in fixed satellite services demand:

- **Growth in fiber optic cable capacity.** Competing telecommunications providers are continually installing new fiber-optic cable between continents, eliminating the need for some existing satellite capacity. Officials from three of the major satellite operators said that they are seeing some demand for satellite services erode with the installation of terrestrial and undersea fiber-optic cables. In its latest competition report, FCC also said that satellite users substitute new fiber-optic transmission cables for satellite services.[24]
- **Improvements in video compression technologies.** Industry officials told us that more efficient video compression technologies allow video to be transmitted with less bandwidth, decreasing overall satellite capacity demand. For example, one network broadcaster told us that more efficient video compression technologies require significantly less satellite capacity than earlier technologies without compression.

According to an estimate by Futron, global demand for fixed satellite services will continue to grow, increasing at an average rate of about 4 percent annually from 2010 through 2019 (from about 7,000 to nearly 10,000 transponder equivalents (TPE)).[25] Consistent with our findings on factors increasing demand for fixed satellite services, Futron projects that, from 2009 through 2019, demand for satellite-based Internet access, enterprise networks, satellite television, and military satellite services will grow faster than the projected industrywide average demand growth rate of 4 percent per year. Specifically, demand is forecasted to grow at rates ranging from 4.7 percent for military satellite services to 10.8 percent for direct satellite-based Internet access services (see fig. 4).

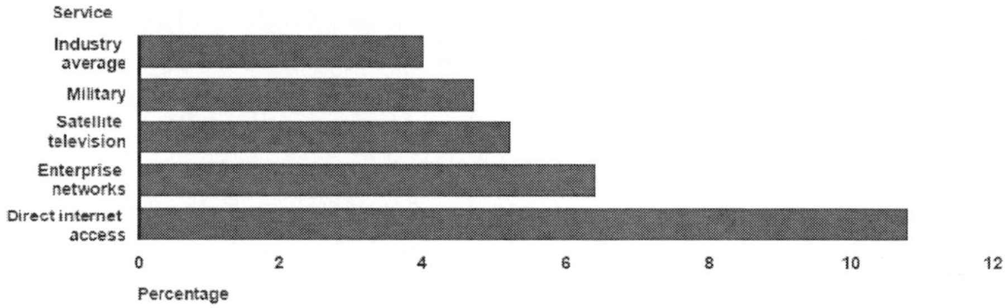

Source: Futron.

Note: Some of the fixed satellite services with zero or negative forecasted growth include video distribution and contribution as well as Internet backhaul. Video distribution is the transmission of entertainment and news content, such as a broadcast network's transmission of video to affiliated stations and cable operators. Video contribution offers point-to-point capacity to providers of media services, such as for video transmission of a news event for later broadcast. Backhaul refers to the transmission of information—or data—from any of a company's aggregation points to an Internet backbone provider that will then transmit the data to any point on the Internet. The growth rates reported above are compounded annually.

Figure 4. Forecasted Percentage Annual Growth in Demand for Fixed Satellite Services.

According to our interviews with industry stakeholders, demand for fixed satellite services is expected to vary and grow at a faster rate in emerging markets with higher rates of economic growth, such as South America and Asia, than in other markets. Futron also forecasts that demand for fixed satellite services in regions with emerging economies will grow as a portion of overall global demand for fixed satellite services from 2009 through 2019. Specifically, Futron expects increases in the share of global demand for fixed satellite services of 5 percentage points for South America, 2 percentage points for South Asia, and 1 percentage point for Sub-Saharan Africa. In contrast, North America's share of worldwide demand for fixed satellite services is projected to decline by 4 percentage points and Western Europe's share is projected to decline by 2 percentage points from 2009 through 2019.

Although expected to grow, according to Futron's estimates, global demand for fixed satellite services will not exceed global fixed satellite capacity, which is also expected to continue growing until 2014. Futron officials told us that capacity forecasts beyond 3 years are difficult to estimate because satellite operators announce satellites expected to be launched only 3 years into the future. Futron said that their estimates beyond 2014 include capacity from new satellites that it expects are likely to replace satellites that are currently in heavy use and nearing the end of their useful lives, but the Futron forecast does not assume the replacement of all satellites currently in orbit. Even given these conservative assumptions, global demand is not expected to exceed global capacity, according to Futron's estimates.

Satellite operators have in the past met and may in the future meet anticipated demand using a variety of strategies.[26] Based on our interviews with stakeholders, the following strategies can be used to increase capacity globally or in specific regions:

- **Launch additional satellite capacity**. As existing satellites reach the end of their useful lives, new satellites have to be launched to replace them and to provide additional capacity to meet any growth in demand for fixed satellite services. Satellite operators told us that they have increased the capacity of their satellite fleets by building and launching satellites with more transponders. According to the Federal Aviation Administration's 2011 commercial space transportation forecast, the average number of transponders on each satellite has increased from 27 in 1993 to 48 in 2011.[27]
- **Move satellites**. Satellite operators told us that they can respond to higher demand and increase capacity in one region by moving existing satellites from another region. For example, two major satellite operators told us that they have moved satellites from one orbital position to another to meet growing demand in other regions.
- **Use hosted payloads**. A hosted payload allows users, such as the government, to add transponders or other equipment to a commercial satellite already scheduled for launch, reducing the time and cost needed to meet demand for satellite capacity, particularly in the short term, assuming resources for hosted payloads can be aligned with satellite manufacturing and launch schedules. Satellite operators said that they are seeing more demand for military use of commercial satellites and hosted payloads as well as increasing demand for hosted payloads from civilian government customers.[28]

INTEGRATION IN THE FIXED SATELLITE SERVICES INDUSTRY HAS ALTERED RELATIONSHIPS BETWEEN OPERATORS AND SERVICE PROVIDERS

Industry Integration since 2000

Since 2000, some fixed satellite operators have merged, and therefore the number of satellite operators providing service to the United States has decreased. Two companies—Intelsat and SES—have emerged as the primary operators of fixed satellite services for the U.S. market, compared with six companies that previously served this market.[29] As figure 5 shows, Intelsat acquired Loral's North American satellites, Comsat General Corporation, and PanAmSat; SES acquired GE Americom, Columbia Communications Corporation, and New Skies. These mergers not only increased the number of satellites in Intelsat's and SES's fleets, but also gave the companies access to the U.S. domestic fixed satellite services market. For example, Intelsat's acquisition of Loral's North American satellites in 2004 gave Intelsat access to U.S. video distribution and corporate data markets; previously, Intelsat offered virtually no U.S. domestic services.[30] Similarly, SES's acquisition of GE Americom, Columbia Communications Corporation, and New Skies combined a predominantly international fixed satellite service operator with domestic satellite providers and allowed SES to enter the U.S. fixed satellite services industry. In some instances, such as Intelsat's acquisition of PanAmSat, the mergers involved satellite operators that served overlapping geographic areas (*a form of horizontal integration*).

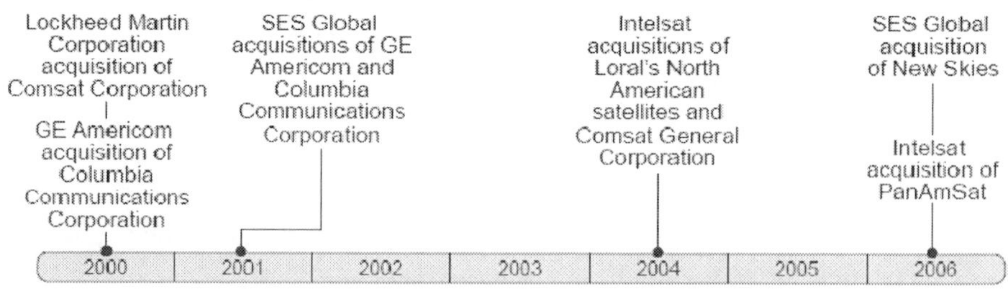

Source: GAO analysis of Federal Communications Commission reports.

Figure 5. Timeline of Mergers among Satellite Operators Providing Fixed Satellite Services to the United States since 2000.

Besides acquiring competitors and potential competitors, both Intelsat and SES have specialized U.S.-based subsidiaries in order to provide direct sales and marketing services to the U.S. government.[31] For example, Intelsat created Intelsat General Corporation through its acquisition of Comsat General, and SES created SES World Skies, U.S. Government Solutions, after it acquired GE Americom.[32] The creation of these subsidiaries (*a form of vertical integration*), combined with the government's adoption of new contracting vehicles, has made more service options and providers available for the U.S. government to procure fixed satellite services bandwidth capacity. Specifically, the FCSA program allows satellite operators, through their subsidiaries, to interact directly with the U.S. government to provide bandwidth capacity, rather than indirectly through a satellite reseller or integrator. Under DOD's former primary contract, DSTS-G, three companies were able to contract with DOD to provide commercial satellite services. Therefore, if DOD needed bandwidth capacity, it would often go through one of these three companies, which would acquire the bandwidth from a satellite operator and resell it to the government.[33] Under the new FCSA program, SES and Intelsat are able to compete through their U.S.-based subsidiaries with resellers and integrators for DOD task orders[34] (see table 1).

Table 1. DSTS-G and FCSA Eligible Vendors as of July 1, 2011

DSTS-G eligible vendors	FCSA eligible vendors
Artel	Maritime Telecommunications Networks
CapRock Government Solutions	Telecommunication Systems
DRS Technical Services	Hughes Network Systems
	Intelsat General Corporation
	CapRock Government Solutions
	Artel
	Americom Government Services
	DRS Technical Services
	Segovia
	Globecomm Systems

Source: GSA.

Note: The list of eligible vendors applies only to the transponded capacity portion of the FCSA program and does not include eligible vendors for the subscription services portion of the program.

Changing Industry Structure and the Relationship between Operators and Service Providers

The horizontal and vertical integration that has occurred in the fixed satellite services industry has changed the structure of the industry and altered the relationship between satellite operators and service providers. Some satellite service providers assert that these structural changes place service providers at a competitive disadvantage;[35] however, to the extent that satellite operators' actions pose anticompetitive concerns, those actions are reserved for review by DOJ and FCC and are outside the scope of this report. According to economic literature we reviewed, as well as interviews with two economists and officials from DOJ's Antitrust Division, these structural changes may also have implications for the end customer, including the U.S. government.

Horizontal Integration

According to economic literature, horizontal integration can increase the market power of the firms remaining after a merger. In the fixed satellite services industry, the mergers that have occurred since 2000 have reduced the number of satellite operators from which large commercial customers, such as broadcasters, and satellite service providers can acquire satellite capacity. To the extent that the satellite operators served the same geographic market, this reduction in the number of operators could therefore lead to increased market power for the remaining companies and higher prices for the customer. For example, one economist told us that with fewer competitors, firms can more easily raise prices, either collectively or individually. In its review of the IntelsatPanAmSat merger, FCC noted that the bargaining power of a buyer of satellite capacity might be reduced as a result of a post-merger reduction in the number of independent, competing satellite operators. However, FCC also noted that, given the size and scale of buyers of satellite capacity, the merger was unlikely to induce any significant adverse effects.[36] Similarly, officials from DOJ, which reviewed the merger, told us that although Intelsat and PanAmSat competed in the same geographic markets, there was insufficient evidence for the agency to prove in court that competitive harm would result from the merger to customers that relied on these companies for service. In particular, DOJ officials told us that at the time of the agency's review, a number of satellite launches were planned that would expand the amount of capacity available to customers. Given the number of alternatives available, DOJ closed its investigation and did not bring a suit in federal court.

While horizontal integration can lead to greater market power and potentially higher prices, it can also have beneficial effects for both the remaining firms and their customers. For example, according to economic literature we reviewed, horizontal integration can lead to economies of scale and efficiencies for the remaining firms as well as lower prices for the customer. One economist also told us that horizontal mergers are often pro-competitive, in that the mergers create efficiencies. In its competition reports, FCC notes that such mergers are beneficial for satellite operators because they can broaden opportunities for achieving economies of scale and scope as operators increase in size. These economies of scale and scope can lead to lower costs for the merged company. In addition, the mergers allowed satellite operators to interact more directly with a broader range of customers and increase the amount of capacity available to serve their customers. For example, by acquiring Loral and PanAmSat, Intelsat not only increased the size of its fleet and therefore its bandwidth capacity serving North America, but it also gained access to new customer markets that it previously served indirectly. Horizontal integration can also benefit customers if the economies and efficiencies operators gain lead to cost savings that are passed on to customers through lower prices.

Vertical Integration

The acquisition of subsidiaries by Intelsat and SES has altered the operators' relationship with satellite service providers in that the subsidiaries are now competing directly with Intelsat's and SES's customers, the service providers, for government contracts. According to one economist, this could allow the operator to favor its subsidiary by charging the satellite service provider a higher price for capacity than it charges its subsidiary. In competing for government task orders, the subsidiary might then be able to submit a lower bid than the service provider, which could place the service provider at a competitive disadvantage. Service providers we spoke with also expressed concern that competing against the subsidiaries of satellite operators could create an uneven playing field in competing for government contracts if the satellite operator charges them a higher price for bandwidth than it charges its own subsidiary. However, if a service provider can acquire capacity from another satellite operator or a terrestrial fiber company, the ability of the satellite operator to disadvantage the service provider is reduced, according to one economist. Additionally, the service provider may be able to compete with the subsidiary on factors other than price. For example, according to satellite operators we spoke with, if a satellite service provider can provide value-added services to the government, it will continue to win task orders, even if the satellite operator can potentially provide lower prices.

In some instances, vertical integration can have beneficial effects for the firms and consumers. According to economic literature we reviewed, a vertically integrated company might be able to provide more attractive service at a lower cost than two firms. Similarly, as DOJ officials noted, a wholesale (or upstream) business moving into the retail (or downstream) business may be able to provide a service more efficiently than two nonintegrated companies providing the same service, and the greater efficiency could ultimately lead to lower prices for the end customer. For this reason, one economist told us, vertical integration may be the best structure for the fixed satellite services industry. In addition, if firms in both the upstream and downstream segments of an industry possess market power, the firms have an incentive to vertically integrate and charge a lower price to the end-user customer. Consistent with this position, one economist noted that higher consumer prices should not result from vertical integration in the fixed satellite services industry; rather, the important concern is maintaining competition at the wholesale, or satellite operator, level. In fact, DOJ officials stated that an industry is generally no more competitive than the least competitive component of the vertical chain. For example, if the industry consists of 10 resellers and 2 operators, the industry's competitiveness is ultimately driven by the competitiveness at the operator level.

Limited Orbital Locations and Enforcement Mechanisms within International Regulations Constrain New Entry into the Industry

Limited Orbital Locations

The number of fixed satellite service satellites operating in geostationary orbit is limited by the availability of orbital locations and the frequencies in which the satellites operate. Although satellites operating at different frequency bands can occupy nearly the same orbital location, roughly 2 degrees of separation is needed between satellites providing coverage in the same geographic area and operating in the same frequency band to avoid interference from neighboring satellites.[37] Certain portions of the geostationary arc are crowded, particularly in the C- and Ku-frequency bands serving certain regions, limiting the potential for new satellites to be launched to serve those regions. For example, portions of the geostationary arc over North America in these frequency bands are largely occupied, according to ITU and satellite stakeholders. Furthermore, once a satellite operator has access to an orbital location, it is unlikely to relinquish that location for use by another operator unless the orbital location is no longer needed to meet demand. According to DOJ officials, the largest constraint for new entry in the fixed satellite services industry is the limited number of satellites that can exist in orbit. Many orbital locations are already taken by existing operators, making it unlikely that other large global satellite operators will enter the industry.

In addition to space and frequency limitations, fixed satellite service satellites are often designed to serve a particular market, service application, or region, which limits their ability to provide service to some customers and may also deter new entrants from serving a particular customer market. That is, although a particular satellite may be capable of serving multiple regions or operating in multiple frequencies, the satellite may direct its signals to serve only a portion of the satellite's overall footprint or the satellite may be operating in only one of its frequency bands. For example, some satellites are launched exclusively to provide satellite television service to the home, and any available capacity on the satellite is usually not leased for other purposes. In addition, an industry official with regulatory expertise told us that satellite operators may not use all the frequencies on their satellites because they choose to reserve capacity. According to a satellite industry consultant, although a new entrant could technically launch a satellite in the same orbital location as an existing satellite to provide coverage in an area or frequency that is not currently being provided, it is unlikely to do so given the high cost of launching a new satellite that could potentially interfere with another operator's satellite in the future.

Regulation and Enforcement

ITU coordinates the use of orbital resources—both orbital locations and frequency assignments—and its Radio Regulations were established and agreed to by member administrations (countries) to ensure equitable and efficient access to these limited resources. As defined in the Radio Regulations, administrations have 7 years from filing an advanced publication to bring a satellite into use.[38] During the 7-year period, administrations must complete a three-stage process:

- **Advanced Publication Information:** An administration must provide a general description of the planned satellite system in order for other administrations to determine the potential effect of this system on already existing systems or other planned systems. This stage triggers the 7-year time frame for bringing a satellite into use and must be completed within 2 years.[39]
- **Coordination:** No earlier than 6 months after providing the information for advanced publication, an administration is required to initiate a coordination procedure, which is a formal regulatory obligation both for the administration seeking a frequency assignment to its satellite network and for the administration whose existing or planned system may be affected by that assignment.[40] Specifically, the administration seeking the frequency assignment is placed in the coordination queue and is required to coordinate with any other administration that has previously initiated the coordination procedure.
- **Notification:** No later than 7 years after filing the advanced publication information, an administration must notify ITU of the date for bringing into use the frequency assignment, to be recorded in the master register.[41] The administration is also required to submit due diligence information on the identity of the satellite network, the satellite manufacturer, and the launch services provider.[42]

According to ITU, once a frequency assignment has been recorded, ITU assumes that the corresponding satellite will operate on a regular basis. However, the frequency assignment may be suspended for a period of 2 years, during which time the administration must either bring the satellite back into operation or replace the satellite.[43] If this does not occur, the administration loses the orbital assignment, which is then removed from the master register and is no longer entitled to recognition. The orbital location then becomes available to any administration in the queue.

According to ITU, the 7-year time frame and coordination procedures were established to ensure that satellite networks are brought into use in a reasonable amount of time, to prevent harmful interference, and to discourage administrations and satellite operators from filing for assignments that they are unlikely to use. For example, if any of the interim milestones are not met and the satellite has not been recorded in the master register within 7 years, the filing is suppressed and the administration loses its position in the queue. According to ITU, the coordination procedures are also meant to facilitate entry into the industry by allowing a satellite network that does not complete the coordination requirements within the 7-year period to receive a provisional recording in the master register.[44] However, even with these efforts, government officials and industry stakeholders identified several factors that may limit entry into the fixed satellite services industry and result in the inefficient use of orbital resources.

- **Excessive filings:** According to ITU and other satellite stakeholders, some administrations file numerous applications for various orbital locations and frequency assignments in order to reserve their place in the queue ahead of other administrations. As a result, other administrations filing for the same orbital location and frequency assignment are required to coordinate with the administrations that have previously filed to avoid interference between the eventual satellite networks. According to ITU, meeting the coordination requirements is time-consuming, costly, and sometimes unsuccessful. In addition, in some cases, administrations are required to coordinate with administrations that have previously submitted filings that are unlikely to actually result in the launch of a satellite. According to ITU, about 20 percent of filings in the advance publication stage reach the notification stage in the 7-year period. According to State Department officials, the requirement to coordinate frequency assignments can act as a deterrent to new entry for both incumbent satellite operators and would-be entrants: An incumbent operator does not have an incentive to coordinate, particularly if the operator has to make costly modifications to an existing satellite in order to avoid interfering with the new satellite. And if the incumbent operator does not coordinate, the would-be new entrant may not be able to obtain financing or may not want to make the large investment needed to construct a satellite that may not operate fully because of interference from an existing satellite.

  Although administrations file numerous applications for orbital locations and frequencies, officials from one global satellite operator told us that this is necessary as they develop their business plans for future satellite launches. Given the cost of constructing a new satellite as well as the time it takes to build and launch a satellite, operators file for orbital locations and frequencies that they expect to need 7 years into the future. During this time, new technologies may be introduced that can alter the business plans of the satellite operator so that it no longer needs the orbital location for which it submitted a filing. However, without the filing, the satellite operator runs the risk of not being able to launch a satellite it has spent time and money developing.

- **Difficulties in interpreting rules:** The definition of "bringing into use" required in the notification stage is unclear and is not consistently interpreted by administrations, according to ITU and other satellite stakeholders. For example, when a satellite is brought into use, it may not have the technical capability to operate at the frequency bands or over the service areas recorded in the master register. In addition, according to one satellite service provider, satellite operators might turn on a satellite for only a short period of time instead of providing consistent operations or move a satellite in inclined orbit into an orbital location as a way to assert ownership over a particular orbital location.[45] These practices could result in the warehousing of orbital resources and make it difficult for new operators to enter the industry, since orbital locations or frequencies that are not in operation are recorded as used in the master register and are therefore unavailable to new entrants. According to ITU and other satellite stakeholders, a clearer definition of bringing into use is needed to ensure that orbital resources are being effectively used.
- **ITU's lack of enforcement authority:** Once an administration notifies ITU that a satellite has been brought into use and the satellite is recorded in the master register, ITU takes the word of the administration that an operational satellite exists in that orbital location and frequency. ITU has no independent way of verifying that a satellite has been launched and is operational. In addition, once a satellite has been launched, ITU does not have a direct way of determining whether the satellite is operating in all of the frequencies in which it was intended to operate or whether the operation has subsequently been suspended. For example, several satellite stakeholders told us that although administrations are required to report unused frequencies and satellites in suspended service to ITU, such reporting rarely occurs, and ITU does not have the means to ensure that a satellite is operating as specified. The Radio Regulations also do not require administrations to submit any specific proof to ITU that they have complied with the due diligence requirements for bringing a satellite into use. The lack of enforcement can hinder entry by not freeing up unused or underused orbital resources.

Given that ITU's Radio Regulations is an international agreement by the member administrations, any changes to the regulations that might improve the use of orbital resources will require the consensus of all member administrations. Although ITU lacks the authority to monitor and enforce the regulations, ITU has taken some steps to address these issues by more aggressively following up with administrations to verify the existence of some satellite networks. For example, since 2009, ITU has taken a more proactive role in identifying satellite networks that are recorded in the master register but do not correspond to an operational satellite, by consulting various databases that track satellites in orbit.[46] ITU contacted the relevant administrations to verify the existence of the satellites, which has resulted in the total or partial suppression of about 100 satellite networks over the last 2 years.[47] These efforts have therefore led to the removal of some unused frequency assignments from the master register, potentially opening these locations and frequencies to other users.

Several ITU conferences and workshops over the last 15 years have also been devoted to improving access to orbital resources. For example, ITU held several workshops from 2008 through 2010 on the efficient use of spectrum and orbital resources. These issues will also be addressed at the upcoming World Radiocommunication Conference, which provides an opportunity for member administrations to amend the Radio Regulations. Several agenda items for the 2012 World Radiocommunication Conference address issues related to fixed satellite services regulations, several of which are issues carried over from previous conferences.[48] For example, one agenda item involves proposals that deal with deficiencies in the advance publication, coordination, or notification and recording procedures for frequency assignments pertaining to space services. According to a satellite industry consultant, proposals under this agenda item would specify a time frame for what constitutes bringing a satellite into use. In addition, the United States has developed several positions for consideration related to this agenda item, including a proposal that would clarify when an administration must notify ITU that a recorded frequency assignment in the master register has been suspended and brought back into use. A separate agenda item addresses any difficulties or inconsistencies encountered by ITU or the member administrations in applying the Radio Regulations. Although the World Radiocommunication Conference provides a forum for administrations to improve the efficiency and equity of orbital resources, broad changes to the regulations are unlikely to occur because consensus is needed from all member administrations, according to a senior ITU official.

Other Factors Affecting Competition

Other factors identified by stakeholders we interviewed that could affect competition in the fixed satellite services industry include the high costs of building, launching, and insuring a satellite and foreign ownership restrictions. Specifically:

- **High costs of entry:** On average, fixed satellite service satellites cost $200 million to $500 million to manufacture, launch, and insure. According to FCC officials, these high fixed costs make it difficult for new operators to enter the fixed satellite services industry, particularly since operators typically need several satellites in orbit to have a basic satellite network. DOJ officials and other satellite stakeholders also stated that the high costs of entry are a barrier for new entry into the fixed satellite services industry.
- **Foreign ownership restrictions:** Various foreign markets have restrictions on foreign ownership that could limit the ability of other companies to provide satellite services in those markets. For example, in China, domestically owned satellite operators receive preferential treatment over foreign satellite operators, according to the Satellite Industry Association.[49] In addition, satellite stakeholders told us that some countries launch satellites for reasons of national pride and do not allow nondomestic operators to use the satellites to provide service.

DOD'S FIXED SATELLITE SERVICE COSTS HAVE INCREASED, BUT CONTRACTING OFFICIALS EXPECT NEW CONTRACT VEHICLES TO INCREASE COMPETITION

Costs since 2003

Using data from DOD and a model that took into account a number of factors thought to influence the cost of fixed satellite bandwidth, we found that the costs to acquire fixed satellite bandwidth increased significantly from 2003 through 2010. Specifically, we acquired procurement data from DOD for its DSTS-G contract, which was DOD's primary mechanism during that time period for acquiring fixed satellite services from commercial vendors. We analyzed the cost, bandwidth purchased, and other attributes of 470 task orders from 2001 through March 2011.[50] In particular, we conducted a regression analysis on the real cost per megahertz (MHz) of bandwidth for each base and option period for the task orders on a number of variables, including frequency band, region, length of task order, option years, and fiscal year in which DISA awarded the task order (see app. II for a complete description of our model). According to our analysis, the real cost per MHz of bandwidth was about 30 percent lower in fiscal year 2003, and lower in all intervening years, than in fiscal year 2010[51] (see fig. 6).

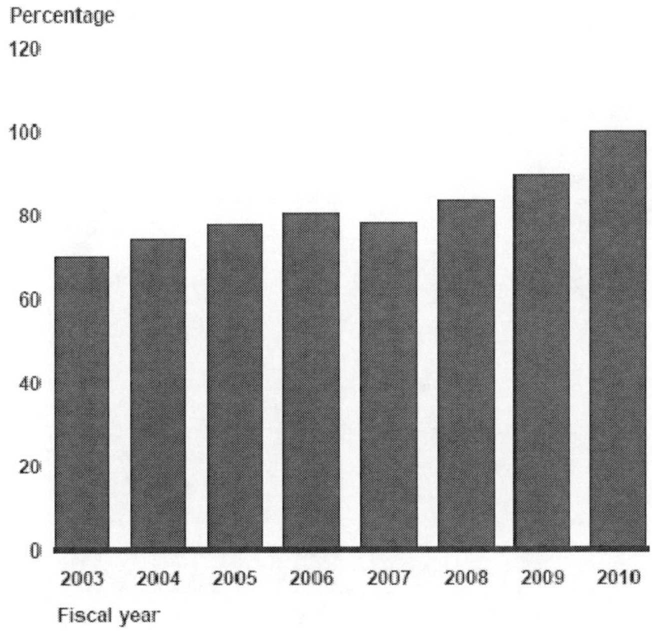

Source: GAO analysis of DOD data.

Note: We derived these estimates from a regression model using data from 2001 through March 2011. We omitted 2001, 2002, and 2011 from this figure as the estimates were not statistically significant at the 5 percent confidence level.

Figure 6. Estimated Real Costs per MHz of Bandwidth for DSTS-G Task Orders as a Percentage of Estimated Costs for Fiscal Year 2010.

Similarly, DOD's own analysis found that bandwidth costs have increased since 2005. Annually, the United States Strategic Command (USSTRATCOM) prepares a report that documents DOD's commercial bandwidth usage and expenditures, including DOD's costs to acquire fixed satellite services; the report provides information on the DSTS-G contract, which accounted for more than 58 percent of DOD's commercial bandwidth expenditures in fiscal year 2009. USSTRATCOM's latest report examined DSTS-G task orders active in fiscal year 2009, with the task orders segmented by the year of award. USSTRATCOM found that the average transponder equivalent (TPE) costs for these task orders increased from $1.1 million in 2005 to $2.5 million in 2009, a $1.4 million (or 127 percent) increase in fixed satellite bandwidth costs over a 4-year period when taking into account the fiscal year in which the task order was awarded[52] (see fig. 7). Thus, while USSTRATCOM uses a different approach from ours to analyze bandwidth costs, both analyses show that costs for fixed satellite services on the DSTS-G contract have increased.

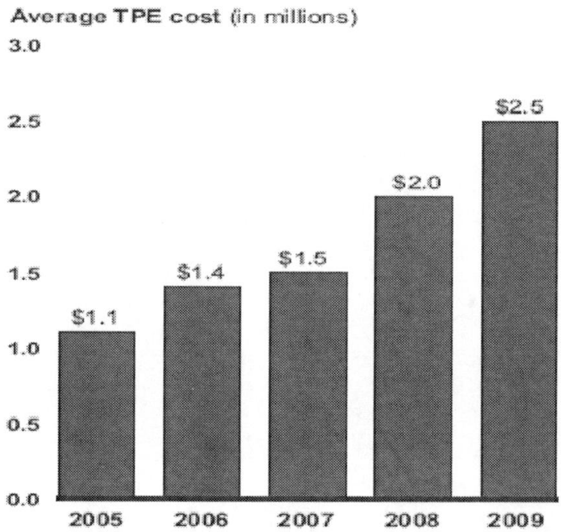

Source: U.S. Strategic Command, Fiscal Year 2009 Commercial Satellite Communications Usage Report, June 9, 2011.

Figure 7. Average Transponder Equivalent Costs of DSTS-G Task Orders Active in Fiscal Year 2009 by Award Year.

Factors Affecting Costs

Integration in the fixed satellite services industry since 2000 has not affected the government's costs to acquire fixed satellite services bandwidth, according to DISA and GSA officials.[53] Government and industry officials maintained that other market factors, such as demand and availability of bandwidth and the length of a task order, have more effect on price. Our analysis of DSTS-G task orders showed that exercising option years resulted in lower bandwidth costs, while longer task orders often resulted in higher costs. In particular, we found that for each year an option is exercised, the cost per MHz of bandwidth decreases by 4.9 percent compared with the base year cost. Similarly, each additional 1 percent increase in the number of days in a task order increases the cost per MHz by 1 percent.[54] Our analysis also found that the cost per MHz was higher in many regions, including Europe and the Middle East, that we analyzed compared with task orders that originated or terminated in North America.[55] These results are consistent with USSTRATCOM's finding that many factors affect bandwidth pricing, such as the amount of bandwidth purchased, the length of the contract, the specific frequency band purchased, the geographic location of satellite coverage, and bandwidth capacity and demand. For example, USSTRATCOM attributed the average TPE cost increase from 2005 through 2009 to the expiration of older task orders that were awarded when more capacity was available and market prices were lower. Furthermore, fewer launches by satellite operators in recent years compared with the late 1990s, along with greater demand for capacity in certain regions, have increased the prices for fixed satellite capacity. Officials from DOJ's Antitrust Division said that the agency did not challenge the Intelsat-PanAmSat or SES-New Skies mergers, partly because the agency thought it could not prove in court that these mergers would negatively affect the government's ability and costs to acquire fixed satellite services.

DISA, GSA, and industry officials expect the new FCSA program to increase competition among eligible vendors and allow the government to contract directly with satellite operators, which may exert downward pressure on the government's cost to acquire fixed satellite services. As previously discussed, three vendors, all satellite service providers, were eligible to compete for fixed satellite services task orders under DOD's DSTS-G contract. With the FCSA transponded capacity and subscription services contracting vehicles, there is no limit to how many eligible vendors, including satellite operators, can compete directly for fixed satellite services task orders. As of July 1, 2011, there were 10 eligible transponded capacity vendors. The increased number of vendors eligible to compete for fixed satellite bandwidth task orders will theoretically exert downward pressure on the cost the government pays to acquire fixed satellite services, assuming the additional vendors can provide a greater number of solutions for a given requirement. In addition, the government's ability to contract directly with a satellite operator for bandwidth capacity when additional services are not needed might exert downward pressure on the cost of satellite bandwidth capacity, since the government can avoid the markup arising from multiple companies providing a single service. GSA officials expect that competition among the set of eligible vendors will exert a downward force on prices, but will be just another factor affecting prices, along with supply and demand and the length of the bandwidth lease. Officials from DOJ's Antitrust Division told us they recognize that the DSTS-G contract limited the number of competitors for the government's fixed satellite services, and said there is reason to think that expanding the number of competitors under the FCSA contracts could result in lower costs.

DISA officials conducted a preliminary analysis of the new transponded capacity and subscription services task orders under FCSA and found that prices are similar to prevailing market prices in recent years and are generally consistent with bandwidth prices awarded under DSTS-G during fiscal years 2009 and 2010. DISA conducted this analysis in response to a recent press article that costs for task orders awarded under the transponded capacity and subscription services were higher than costs under DSTS-G. DISA based its analysis on 35 task orders that have been awarded through March 2011. DISA officials told us they will continue to analyze prices for transponded capacity and subscription services and are optimistic that prices will remain competitive given the number of vendors that are able to bid on the task orders.

Agency Comments

We provided a draft of this report to DOD, DOJ, FCC, GSA, and the Department of State. The agencies provided technical comments that we incorporated as appropriate.

Mark L. Goldstein
Director, Physical Infrastructure

Appendix I: Objectives, Scope, and Methodology

This report examines the state of competition in the fixed satellite services industry. In particular, the report describes: (1) how current satellite capacity compares to current and forecasted demand; (2) the changes that have occurred in the fixed satellite services industry since 2000 and the effects these changes could have on the relationship between satellite operators and service providers; (3) technological, regulatory, and other factors that affect competition in the fixed satellite services industry; and (4) how costs for the Department of Defense (DOD) to acquire fixed satellite services have changed since 2000 and contracting officials' views on the effects of changes in the industry and contracts on costs.

To determine how current fixed satellite capacity compares to current and forecasted demand, we used demand and capacity data contained in the Executive Summary of Futron's *2010 Forecast of Global Satellite Services Demand* report, which analyzed trends in the industry from 2009 through 2019. We took steps to ensure the reliability of the data in the Futron report, including determining the procedures and controls that Futron used to make estimates about future capacity and demand for fixed satellite services. We determined that the data were sufficiently reliable for the purposes of our report. We interviewed officials from three research firms that analyze demand and capacity in the fixed satellite services industry—Euroconsult, Futron, and Northern Sky Research—as well as satellite operators, service providers, and industry analysts, about factors affecting demand for fixed satellite services. We reviewed the Federal Aviation Administration's *2011 Commercial Space Transportation Forecasts* report showing yearly estimates of commercial satellite launches through 2019 and the Satellite Industry Association's 2011 *State of the Satellite Industry Report* prepared by Futron. We reviewed financial reports from the four major satellite operators—Intelsat, Eutelsat, SES, and Telesat—for statements on current fixed satellite capacity and demand and demand for enterprise networks.

To determine the changes that have occurred in the fixed satellite services industry since 2000, and the effects these changes could have on the relationship between satellite operators and service providers, we reviewed the Federal Communications Commission (FCC) reports to Congress on the status of competition in the markets for domestic and international satellite communications services and annual reports as required by federal law, as well as comments filed by industry stakeholders for these reports.[1] We conducted an economic literature review of horizontal and vertical integration in the satellite and related telecommunications industries and interviewed economists with expertise in horizontal and vertical integration about the factors that would lead companies to integrate and the advantages and disadvantages of such integration for competition. In addition, we interviewed officials from the

Department of Justice's Antitrust Division, which reviewed mergers in the fixed satellite services industry, about the department's process for reviewing potential mergers and the potential implications of horizontal and vertical integration for competition in the fixed satellite services industry.

To determine the technological, regulatory, and other factors that affect competition in the fixed satellite services industry, we reviewed FCC's competition reports to further understand the regulatory factors that could affect competition and interviewed officials from FCC's International Bureau about its process for licensing fixed satellites and the effects of the International Telecommunication Union's (ITU) regulations on competition. ITU is an international organization formed to promote international cooperation and develop consensus positions within the telecommunications and information technology industries. We reviewed information on ITU's Radio Regulations governing the use of orbital resources; the regulations are agreed to by the member administrations (countries) of ITU. We interviewed a senior ITU official about the effect of these regulations on competition in the fixed satellite services industry and obtained additional documentation about proposals to modify these regulations to improve the efficient and equitable use of orbital resources. We interviewed officials from the Department of State, two industry officials with regulatory expertise identified through our interviews, and satellite industry stakeholders about the regulatory factors that could affect competition.

To determine how costs for DOD to acquire fixed satellite services have changed since 2000, and to obtain contracting officials' views on the effect of changes in the industry and contracts on the cost for the government to acquire fixed satellite services, we obtained data from the Defense Information Systems Agency's (DISA) Annual Usage Report Database on total fixed satellite services expenditures and bandwidth costs across various contracts DOD uses to procure commercial satellite services.[2] We limited our analysis to task orders under the department's Defense Information Systems Network Satellite Transmission Services-Global (DSTS-G) contract because it is the primary contract DOD uses to acquire fixed satellite services, accounting for more than 58 percent of DOD's fixed satellite services expenditures in 2009. In particular, we analyzed bandwidth costs on 470 DSTS-G task orders awarded from 2001 through March 2011 by regressing the real cost per megahertz of bandwidth on a number of variables thought to influence bandwidth costs, including frequency band, region, length of the task order, options years, and fiscal year in which DISA awarded the task order, to determine whether costs had increased over time[3] (see app. II for a detailed description of our model). We obtained a tutorial on the database from DISA officials responsible for maintaining the database to determine how the data are collected, what controls are in place to ensure the reliability of the data, and any limitations to the data. We determined that the data were sufficiently reliable for the purposes of our review. We compared the results of our analysis to DISA's own analysis of bandwidth costs using the same data set. Specifically, we reviewed spend analysis reports prepared by DOD, as required by federal law, on commercial satellite communications services used by the department, and annual usage reports prepared by the United States Strategic Command with assistance from DISA on the department's use of and expenditures for commercial satellite communications services.[4] These reports include DOD analyses on the costs to acquire fixed satellites services on the DSTS-G contract calculated on a transponder equivalent (TPE) basis, where one TPE is equal to one 36-megahertz transponder for a duration of 1 year. We interviewed DISA and General Services Administration (GSA) officials to obtain their views on how changes in the industry have affected the cost for the government to acquire fixed satellite services and how new contract vehicles developed by DISA and GSA could affect costs. We also interviewed officials from three major broadcast companies with responsibility for procuring commercial satellite bandwidth for their programming in order to obtain the commercial perspective on costs compared to the government perspective. We selected the companies based on the number of cable networks owned, whether the company had affiliated television broadcast networks, and the prime-time ratings of the cable networks.

Appendix II: Regression Model and Analysis of the Costs DOD Incurs to Acquire Fixed Satellite Services

This appendix describes the model we developed to analyze the costs that DOD incurs to acquire fixed satellite services from commercial vendors. Specifically, we discuss (1) background information on the factors that affect the cost to acquire fixed satellite services and DOD's analyses, (2) the data we used in our analysis, and (3) our estimation methodology and results.

Background

DISA acquires fixed satellite services for DOD from commercial vendors. According to DISA officials, a variety of factors affect the cost to acquire satellite bandwidth, including a number of intangible factors. For example, DISA officials noted that option years on a base-year contract are exercised 90 to 95 percent of the time, which increases their value to industry. Furthermore, DISA officials said that the fill rate—the percentage of a satellite's capacity utilized—plays a role in price; namely, higher fill rates, which imply less available capacity, are associated with higher bandwidth costs. Private equity ownership of Intelsat and other satellite operators is influencing the number of satellite launches occurring and the fill rates of those satellites.

Annually, the United States Strategic Command (USSTRATCOM) prepares a report that documents DOD's commercial bandwidth usage and expenditures, including DOD's costs to acquire fixed satellite services on the DSTS-G contract, which accounted for more than 58 percent of DOD's commercial bandwidth expenditures in fiscal year 2009. DISA obtains information on DOD bandwidth expenditures by surveying its departments using a standard data collection template. The data are reviewed for accuracy and completeness, and comparative analyses are conducted across various factors, including region, frequency band, satellite operator, and customer.[1] In addition, bandwidth costs are normalized by converting the total lease cost to an annualized 36- megahertz TPE cost for comparison purposes. USSTRATCOM's latest report found that DOD's average TPE costs increased from $1.1 million in 2005 to $2.5 million in 2009, a $1.4 million (or 127 percent) increase in fixed satellite bandwidth costs over a 4-year period when taking into account the fiscal year in which the task order was awarded.[2]

Data Source and Descriptive Statistics

To conduct our analysis, we acquired procurement data from DISA for its acquisition of fixed satellite services. The dataset included 527 task orders awarded from fiscal year 2001 through March 2011 for the DSTSG contracting vehicle, which was DOD's primary mechanism during that time period for acquiring fixed satellite services from commercial vendors.[3] Within the dataset, each task order was represented by one or more observations; one observation represented the base period (all task orders had a base period) and one or more additional observations represented option periods (not all task orders had option periods). The dataset included the following information for each observation:

(1) contract number, (2) task order number, (3) start date, (4) end date, (5) number of days covered by an observation, (6) the fiscal year when a task order began, (7) the fiscal years covered by a task order, (8) vendor, (9) satellite operator, (10) satellite number, (11) bandwidth capacity, (12) frequency band, (13) three location fields generally designating regions of the world from which bandwidth was transmitted and received, (14) total cost, (15) bandwidth cost, (16) "other" costs, (17) annualized 36-megahertz (MHz) TPE cost, and (18) a yes/no field indicating whether or not a contract is recurring. In addition, each observation contained this information for the base year in which the task order was awarded as well as any option years that may have been exercised, which are indicated by the fiscal years covered in a task order.

For each task order, we used information on the bandwidth cost, bandwidth capacity (the amount of bandwidth leased), number of days covered by an observation, frequency band, and region where the bandwidth transmission originated and terminated. We used this information for the base period and for each option period of the task order. In addition, we aggregated those observations for a task order that occurred in the same fiscal year in order to obtain yearly observations. Specifically, to aggregate observations within a fiscal year, we:

- computed the duration by ascertaining the difference between the earliest start date and the latest end date;
- summed bandwidth costs, total costs, and other costs to derive the aggregate costs across observations for the fiscal year;
- derived a "weighted MHz" by computing (1) the number of days covered by an observation and dividing it by the duration; (2) multiplying the respective bandwidth capacity by the number resulting from (1); and (3) summing the numbers resulting from (2); and
- ascertained whether a task order so aggregated into a fiscal year encompassed varying frequency bands or regions and, if it did, recoding those fields as "Multiple" to reflect this multiplicity.

Once we completed the aggregation process, we eliminated those task orders that had no bandwidth costs and verified any anomalies with DISA. This resulted in a total of 470 task orders for our analysis. The resulting dataset is considered a panel dataset in which task orders are tracked through time. However, every task order is not present in every year. Hence, the final dataset is an unbalanced panel, since the number of observations for each year is not the same. Finally, we converted the costs into 2010 dollars.

Table 2 presents the yearly descriptive statistics for our dataset. Column 2 shows the average cost per task order by year, deflated using the Producer Price Index for Communications Systems and Equipment, including microwave and space satellites. Column 3 shows the average bandwidth acquired by year. Column 4 presents the average cost per MHz of bandwidth. Column 5 presents the average number of days per task order. Column 6 shows the proportion of task orders that use the Ku-band frequency, which is the most commonly used frequency band; the C-band accounts for most of the remaining frequency. Column 7 shows the proportion of task orders using multiple frequencies; since each task order could comprise several contracts, these independent contracts might use more than one frequency. Finally, column 8 presents the number of task orders per year in our dataset.

Table 2. Descriptive Statistics

Fiscal year	Real cost (2010 dollars)	Bandwidth (MHz)	Average cost per MHz	Days in task order	Ku-band frequency (percent)	Multiple frequency (percent)	Number of task orders
2001	$942,676	12.24	$97,256	304.40	60.00	0.00	5
2002	3,339,774	77.30	64,255	336.91	86.96	0.00	23
2003	2,259,241	52.69	41,459	250.54	75.61	2.44	41
2004	1,700,018	43.47	39,370	249.81	74.32	6.76	74
2005	1,472,920	38.03	36,763	236.16	73.15	5.56	108
2006	1,781,709	44.76	57,859	292.53	74.31	4.59	109
2007	1,674,661	44.69	41,302	261.63	75.78	2.34	128
2008	2,140,518	52.37	50,214	309.98	72.66	3.91	128
2009	2,177,265	51.65	40,775	262.48	73.49	3.01	166
2010	2,733,764	66.06	44,497	268.63	75.19	6.20	129
2011	706,645	60.01	15,643	95.82	71.43	10.71	28

Source: GAO analysis of DOD data.

Estimation Methodology and Results

We modeled the task order cost as a function of a number of variables thought to influence the cost to acquire fixed satellite services, such as the bandwidth acquired, frequency band, and region. As discussed previously, our final dataset is an unbalanced panel, since the number of observations for each year is not the same. In addition, we assumed that there is an unobserved effect within each task order that is independent of the rest of the explanatory variables. Therefore, we estimated our model with a random effect regression. In particular, we estimated the following equation:

$$\ln C_{it} = \alpha + \beta_1 option_{it} + \beta_2 \ln days_{it} + \beta_3 ku_{it} + \beta_4 multi\_freq_{it} + \\ + \delta year\_dummies_{it} + \gamma region\_dummies_{it} + a_i + u_{it}$$

where:

α= intercept

ln C_{it} = the logarithm of the real cost per megahertz of bandwidth for task order i in year t (in 2010 dollars).

$option_{it}$= a variable that indicates if task order i is exercised in option year t. It is the difference between the initial year the task order was awarded and the year the option is exercised.

ln $days_{it}$ = the logarithm of the number of days task order i was active in year t.

kuit = a dummy variable indicating if the frequency for task order i and year t is the Ku-band.

$multi\_freq_{it}$ = a dummy variable indicating if the frequency for task order i and year t is more than one band.

$year\_dummies_{it}$ = a set of dummy variables controlling for each year using 2010 as the reference year.

$region\_dummies_{it}$ = a set of dummy variables controlling for the origination and destination transmission regions using the North America region as the reference region for task order i in year t.

a_i = task order unobserved effect assumed to be uncorrelated with the explanatory variables.

uit = error term assumed to be independent and identically distributed with a normal distribution with zero mean and variance y^2.

The results from our model are shown in table 3. With the random effects panel data estimation, we used fiscal year 2010, C-band, and North America as the reference groups for fiscal year, frequency, and region, respectively.

Table 3. Regression Results—Dependent Variable Logarithm of the Real Cost per Megahertz of Bandwidth

Variable	Coefficient	Standard errors	P-value
Option	-0.049[b]	0.022	0.027
Log(days)	1.000[a]	0.036	0.000
Ku-frequency	-0.312[a]	0.105	0.003
Multi-frequency	-0.187	0.245	0.446
Fiscal year 2001	-0.336	0.237	0.157
Fiscal year 2002	-0.257	0.176	0.143
Fiscal year 2003	-0.300[b]	0.144	0.038
Fiscal year 2004	-0.256[b]	0.131	0.050
Fiscal year 2005	-0.226[b]	0.101	0.024
Fiscal year 2006	-0.197[b]	0.087	0.024
Fiscal year 2007	-0.221[a]	0.068	0.001
Fiscal year 2008	-0.166[a]	0.053	0.002
Fiscal year 2009	-0.104[a]	0.033	0.001
Fiscal year 2011	0.001	0.095	0.991
Asia	0.814[a]	0.208	0.000
Atlantic Ocean region	0.369	0.219	0.092
Europe-Middle East-Africa	0.289[b]	0.115	0.012
Europe	0.430[a]	0.128	0.001
Middle East	0.395[b]	0.174	0.023
Middle East-Africa	0.062	0.099	0.532
Multi-region	0.297[a]	0.077	0.000
Pacific Ocean region	0.237[b]	0.121	0.049
Constant	5.160[a]	0.210	0.000
Number of observations		939	
Number of groups		470	
Overall R-squared		0.822	

Source: GAO analysis of DOD data.

Note: Robust standard errors reported. [a]Significance at the 1 percent level. [b]Significance at the 5 percent level.

Our model results indicate that the cost DOD incurs to acquire fixed satellite services have increased since 2003. Controlling for other variables thought to influence the cost to acquire fixed satellite services, the cost per megahertz was lower in fiscal years 2003 through 2009 than in fiscal year 2010.[4] According to DISA officials, the recent increases in costs across most of the industry have been driven mostly by the greater fiscal discipline of satellite operators, leading to more conservative investments in new satellite launches, compared to the highly speculative and aggressive investing that the industry underwent in the dot-com era of the late 1990s. Fewer and more deliberate launches, coupled with a steady increase in demand, have resulted in recent higher utilization rates and, therefore, higher prices.

In addition to the trend in costs over time, our results illustrate that exercising an option year is associated with lower costs while task orders that last longer are associated with higher costs. In particular, we found that for each year an option is exercised, the cost per megahertz of bandwidth decreases by 4.9 percent compared with the base year cost. DISA officials said that option years on a base-year contract are exercised 90 to 95 percent of the time and therefore, industry treats these contracts as multiyear procurements, which result in lower costs. We also found that each additional 1 percent increase in the number of days in a task order increases the cost per MHz by 1 percent.[5] We found that Ku-band task orders are associated with 31 percent lower costs than C-band task orders. Finally, many regions, including Asia, Europe, the Middle East, and the Pacific Ocean region, have higher costs than North America.

End Notes

[1] Fixed satellite service refers to a radiocommunication service between fixed earth stations at specific locations by means of one or more satellite.

[2] Satellite Industry Association, *State of the Satellite Industry Report* (prepared by Futron Corporation) (Washington, D.C.: June 2011).

[3] Consumer satellite services, including satellite television, accounted for 82 percent of revenues in the satellite services sector. For purposes of its report, the Satellite Industry Association separates consumer satellite services from fixed satellite services. The remaining revenues in the fixed satellite services sector were from managed network services, where a satellite provider delivers voice, data, Internet, and video network services for multi-site enterprises.

[4] United States Strategic Command, *Fiscal Year 2009 Commercial Satellite Communications Usage Report* (June 9, 2011).

[5] Airborne intelligence, surveillance, and reconnaissance systems include manned and unmanned airborne systems, and play a critical role in supporting current and future military operations and national security missions by collecting, processing, and disseminating data.

[6] Pub. L. No. 106-180, 114 Stat. 48 (2000).

[7] Mobile satellite service satellites may also operate in geostationary orbit. Unlike fixed satellite service satellites, which communicate primarily with ground stations located at fixed points, mobile satellite service satellites are primarily used to communicate with mobile devices.

[8] A transponder aboard a communications satellite receives the uplink signal sent from the ground, shifts its frequency to the downlink frequency, amplifies it, and transmits it to the ground.

[9] This signal degradation is referred to as "rain fade." Higher frequency signals are more likely to be susceptible to these effects.

[10] Hybrid satellites are designed to operate in more than one frequency band, for example, both C- and Ku-band transponders on one satellite.

[11] A corporate enterprise network is a communications backbone that connects all of a company's networks, such as its computer systems, at all of its locations.

[12] Satellite television is also referred to as Direct-to-Home television, which provides television service directly to the home. In the United States, DirecTV and DISH Network own and operate their own direct broadcast satellites to provide this service to customers.

[13] Whereas C-band satellites usually cover large areas of the earth's surface with one beam, such as the continental United States, Ka-band satellites have spot beams that may be sized to cover only a metropolitan area.

[14] The FCSA program will also replace the existing Inmarsat mobile satellite services contract; Inmarsat is a mobile satellite services operator. The departments within DOD also procure fixed satellite services through various other contracts. However, DSTS-G is the primary contract through which DOD procures its bandwidth capacity.

[15] The FCSA program created two new Schedule Item Numbers under the GSA Schedule 70 Multiple Awards Schedule for transponded capacity and subscription services. The transponded capacity category is for dedicated satellite bandwidth in any commercially available frequency band, and the subscription services category is for turnkey, pre-engineered subscription-based solutions. FCSA also developed two multiple award Indefinite Delivery/Indefinite Quantity solicitations for end-to-end satellite communication solutions.

[16] The ORBIT Act also provided a U.S. regulatory framework for the privatization of Inmarsat, a mobile satellite services operator.

[17] See, GAO, *Telecommunications: Market Developments in the Global Satellite Services Industry and the Implementation of the ORBIT Act*, GAO-05-550T (Washington, D.C.: Apr. 14, 2005).

[18] 47 U.S.C. § 765e.

[19] 47 U.S.C. § 703.

[20] Futron, *2010 Futron Forecast of Global Satellite Services Demand, Executive Summary* (Bethesda, Md: 2010). All Futron data cited in this report are from this source unless otherwise noted.

[21] The fleet utilization rates for the four major satellite operators as of their March 2011 financial reports were Eutelsat, 91 percent; Intelsat, 78 percent; SES, 80 percent; and Telesat, 83 percent. Futron released its 2010 demand forecast report that estimated a 79 percent capacity utilization for 2011 for all fixed satellite fleets worldwide in November 2010.

[22] FCC, *Second Annual Report and Analysis of Competitive Market Conditions with Respect to Domestic and International Satellite Communications Services* (Washington, D.C.: October 2008).

[23] Satellite Industry Association, *State of the Satellite Industry Report*, June 2011.

[24] FCC, *Second Annual Report*, 2008.

[25] This forecast is Futron's baseline or most likely demand scenario. We did not have access to Futron's low- and high-demand scenarios. A TPE is a traditional industry metric that defines the total capacity on a satellite in terms of transponders with a bandwidth of 36 megahertz.

[26] Strategies to meet anticipated demand have long lead times (e.g., the time from design to launch is anywhere from 7 to 15 years) and might not be responsive to an immediate, unanticipated surge in demand.

[27] Federal Aviation Administration, *2011 Commercial Space Transportation Forecast* (Washington, D.C.: May 2011).

[28] We have reported that planned DOD military satellite programs experiencing cost growth have either been delayed or discontinued. See, for example, GAO, *Space Acquisitions: DOD Faces Substantial Challenges in Developing New Space Systems*, GAO-09-705T (Washington, D.C.: May 20, 2009). DOD officials told us that there are not enough current and projected DOD-owned satellites to meet demand and therefore DOD will have to lease more capacity from commercial satellites.

[29] In addition to Intelsat and SES, Telesat and Eutelsat provide some limited fixed satellite services to the United States.

[30] Before it was privatized in 2001, INTELSAT competed with domestic satellite operators to provide fixed satellite capacity for services to and from the United States.

[31] Intelsat and SES also have subsidiaries to serve other customer groups, such as media companies.

[32] The acquisition of these subsidiaries enabled Intelsat and SES, which are predominantly European-owned, to compete for classified U.S. government business for fixed satellite services. Although the subsidiaries are owned and controlled by non-U.S. parent corporations, they have separate boards and functions from the parent companies. Intelsat and SES entered into proxy agreements, as required by the U.S. government, to protect against foreign ownership, control, and influence.

[33] DOD department are required to acquire satellite service through DISA, and DSTS-G was DISA's primary contract vehicle. However, DOD departments could procure satellite services directly from the U.S. subsidiaries of satellite operators through other contracts outside of DSTS-G. In addition, the U.S. subsidiaries of satellite operators were eligible to compete for non-DOD government contracts under GSA's SATCOM-II contract.

[34] A task order establishes the deliverables and costs and includes a base period—for example, 1 year—and perhaps one or more option periods.

[35] For example, see GAO, *CapRock Government Solutions, Inc.; ARTEL, Inc.; Segovia, Inc.*, B-402490 (Washington, D.C.: May 11, 2010).

[36] FCC, *In the Matter of Constellation, LLC, Carlyle PanAmSat I, LLC, Carlyle PanAmSat II, LLC, PEP PAS, LLC, and PEOP PAS, LLC , Transferors and Intelsat Holdings, Ltd., Transferee, Consolidated Application for Authority to Transfer Control of PanAmSat License Corp. and PanAmSat H-2 Licensee Corp.*, FCC-06-85 (June 19, 2006). According to FCC, mergers and acquisitions can eliminate a market participant and, at the same time, create a more competitive post-merger firm if the depth and breadth of its services are greater than before the merger.

[37] FCC's rules for protection from frequency interference are based on fixed satellite service satellites operating in geostationary orbit at separations of no more than 2 degrees. 47 C.F.R. § 25.140(b)(2).

[38] ITU Radio Regulation No. 11.44. In addition, ITU requires administrations to pay a cost-recovery fee in order to process filings, which vary based on the complexity and size of the filing.

[39] ITU Radio Regulation No. 9.5D.

[40] ITU Radio Regulation No. 9.1.

[41] ITU Radio Regulation No. 11.44.

[42] ITU Resolution 49, Section 2.47.

[43] ITU Radio Regulation No. 11.49.

[44]ITU Radio Regulation No. 11.41. A new satellite network can be provisionally recorded in the master register prior to completing all coordination requirements and receive a definitive recording, if during a period of 4 months, both the newcomer and the existing satellite network have been in operation without any complaints of harmful interference. According to ITU, this provision is intended to facilitate the entry of a newcomer even if another administration was first in line for a particular orbital location.

[45] A satellite in inclined orbit is typically one that is near the end of its useful life and does not have enough fuel to maintain its position over the equator in the north-south direction. Consequently, the satellite oscillates in an orbit inclined above and below the equatorial plane. Such satellites have limited operational capabilities because they typically require the use of specialized ground stations.

[46] ITU has undertaken these actions in accordance with ITU Radio Regulation No. 13.6.

[47] Suppressed filings are removed from the master register.

[48] World Radiocommunication Conferences are held every 3 to 4 years to review and, if necessary, revise the Radio Regulations governing the use of the radio-frequency spectrum. The revisions are based on the agenda developed by ITU, which take into account recommendations made in previous conferences.

[49] *Comments of the Satellite Industry Association to the U.S. Trade Representative for its 2010 1377 Review of Telecommunications Trade Agreements*.

[50] In total, 527 task orders were awarded on the DSTS-G contract from 2001 through March 2011. However, we eliminated task orders that had no bandwidth costs and three outliers identified for which information was not available.

[51] Our estimates of the change in cost were not statistically significant for fiscal years 2001, 2002, and 2011 compared with fiscal year 2010; while our results indicated that the cost per MHz was lower in fiscal years 2001 and 2002, and higher in fiscal year 2011, than in fiscal year 2010, these results were not statistically significant at the 5 percent confidence level.

[52] United States Strategic Command, *Fiscal Year 2009 Commercial Satellite Communications Usage Report*, 2011. For its report, USSTRATCOM calculates the cost of fixed satellite services on a TPE basis, where a TPE is equal to one 36-megahertz transponder for a duration of 1 year. USSTRATCOM also reported that the average TPE cost for all active task orders, which includes task orders awarded in and prior to a fiscal year, increased from $1.1 million in fiscal year 2005 to $1.7 million in fiscal year 2009.

[53] Two of the major broadcast companies we interviewed said that consolidation had not affected the price for fixed satellite service.

[54] DISA officials said that according to their analysis, when measured on a cost per MHz per day basis, task orders with a longer duration have lower costs than shorter duration task orders. In other words, shorter duration task orders have a higher cost per day than longer duration task orders.

[55] The regions included in our analysis were Asia, Europe, North America, Middle East, Middle East-Africa, Europe-Middle East-Africa, Atlantic Ocean, Pacific Ocean, and Multi-regions (combinations of two or more of the above regions).

End Notes for Appendix I

[1] 47 U.S.C. §§ 703, 765e.

[2] We limited our analysis to DOD's costs to acquire fixed satellite services because of limitations on the availability of data.

[3] There were a total of 527 DSTS-G task orders awarded from 2001 through March 2011. However, we eliminated task orders from our analysis that had no bandwidth costs, as well as three task orders that we considered outliers and for which information was not available. We limited our analysis to bandwidth costs because we wanted to assess satellite-related costs as opposed to costs for terrestrial services. In addition, bandwidth costs generally account for about 80 percent of overall task order costs.

[4] See, *Commercial Satellite Communications Service Spend Analysis and Strategy Report in Response to Section 818 of Public Law 109-163 – the National Defense Authorization Act for Fiscal Year 2006* (June 7, 2006); Pub. L. No. 109-163 § 818, 119 Stat. 3136, 3384 (2006); DOD, *Investment and Acquisition Strategy for Commercial Satellite Capabilities Report in Response to Section 912 of Public Law 110-417 – the National Defense Authorization Act for Fiscal Year 2009* (June 3, 2010), Pub. L. No. 110-417, § 912, 122 Stat. 4356, 4571 (2008); and United States Strategic Command, *Fiscal Year 2009 Commercial Satellite Communications Usage Report* (June 9, 2011).

End Notes for Appendix II

[1] To ensure the accuracy and completeness of the data, the annual usage database is validated annually using historical data collected in reports for the past years and the results are reviewed by the United States Strategic Command and DOD departments that acquired commercial satellite communications services for the year.

[2] United States Strategic Command, *Fiscal Year 2009 Commercial Satellite Communications Usage Report*, 2011. USSTRATCOM also reported that the average TPE cost for all active task orders, which includes task orders awarded in and prior to a fiscal year, increased from $1.1 million in fiscal year 2005 to $1.7 million in fiscal year 2009.

[3] A task order establishes the deliverables and costs and includes a base period—for example, 1 year—and perhaps one or more option periods.

[4] The coefficients for fiscal year 2001 and 2002 are consistent with lower costs than fiscal year 2010, and the coefficient for fiscal year 2011 is consistent with higher costs than fiscal year 2010, further illustrating the upward trend in costs. However, these coefficients are not statistically significant.

[5] DISA officials said that according to their analysis, when measured on a cost per MHz per day basis, task orders with a longer duration have lower costs than shorter duration task orders. In other words, shorter duration task orders have a higher cost per day than longer duration task orders.

In: Satellite Communications ISBN: 978-1-61942-464-7
Editors: B. G. Osborn & E. T. Rowland © 2012 Nova Science Publishers, Inc

Chapter 2

NSTAC REPORT TO THE PRESIDENT ON COMMERCIAL COMMUNICATIONS RELIANCE ON THE GLOBAL POSITIONING SYSTEM (GPS)*

President's National Security Telecommunications Advisory Committee

EXECUTIVE SUMMARY

The President's National Security Telecommunications Advisory Committee (NSTAC) performed an evaluation in response to a White House request for commercial communications industry findings on the commercial communications infrastructure's reliance on the Global Positioning System (GPS). To gain current perspectives on the industry-wide use of GPS, the NSTAC solicited information from its members, other providers within the industry, and several external subject matter experts. Specifically, the NSTAC requested information on: (1) company and industry segment use of and

---

* This is an edited, reformatted and augmented version of the President's National Security Telecommunications Advisory Committee publication, dated February 28, 2008.

reliance on GPS signals; (2) impacts to networks and operations that would result from loss or degradation of GPS signals; and (3) specific strategies implemented or planned to mitigate the impact of any GPS signal loss or degradation.

A broad cross-section of the commercial communications industry submitted feedback, including responses from individual companies in the telecommunications, computer software/services, and aerospace and defense sectors, and from industry trade associations. As a result of the evaluation, the NSTAC developed several findings and a recommendation for White House review and consideration.

This evaluation focuses narrowly on the commercial communications industry's use of and reliance on GPS, in accordance with White House direction to tailor the effort to enable a quick response. The NSTAC notes that reliance on GPS signals in military, maritime, aviation, and other civil environments varies widely depending on the specific use and application. GPS-based precision-guided munitions, maritime harbor approach and constricted waterway navigation, and aviation approach and landing are examples of critical applications with varying positioning, navigation, and timing (PNT) requirements supported by GPS. While instructive in understanding overall GPS deployment trends and vulnerabilities, non-commercial information was not evaluated or integrated into the NSTAC's findings, as that area was deemed to be outside the scope of this effort. The NSTAC looks forward to ongoing engagement with the White House staff and offers its continued support to national security telecommunications policy development and program planning.

Study Findings and Recommendation

The U.S. Government's commitment to provide and maintain civil space-based PNT services, such as GPS, free of direct user fees for civil, commercial, and scientific uses has encouraged the rapid adoption of GPS-based solutions throughout the commercial communications industry. In today's environment, GPS supports a broad range of commercial communications industry functions and applications; the primary use of GPS in each industry segment is in support of the networks' precise timing and synchronization requirements. Companies have selected and widely implemented GPS-based solutions primarily because GPS provides an inexpensive, globally-available, and highly reliable Stratum 1-quality reference source. As the commercial communications network infrastructure continues to evolve toward a high-speed all-digital environment, accurate timing and synchronization functions that support the infrastructure are becoming more critical.

Another important use of GPS is support to wireless location-based services, including support of wireless Enhanced 911 (E911) Phase II requirements. As the overall market for GPS-based devices and services continues to grow, the commercial communications industry is likely to identify and utilize additional uses of GPS to increase productivity, service delivery, and the number of available end-user applications.

Because of the fundamental role that GPS plays in supporting the commercial communications infrastructure, industry employs a range of strategies to mitigate the impact of GPS loss or disruption. To protect critical functions such as network timing and synchronization, companies proactively employ multiple layers of backup capabilities, mitigation strategies, and contingency plans to ensure protection against a wide range of potential GPS outage or disruption scenarios. At critical nodes in the infrastructure, redundant Stratum 1-level sources are deployed and protected automatically by secondary and tertiary backup capabilities and alternate timing sources. All major carriers adhere to extremely rigorous industry-standard requirements for network timing synchronization.

Technological, economic, and regulatory considerations necessarily factor into individual company decisions on how to mitigate the potential impact of GPS loss. Companies must consider available equipment types and cost, the required level of quality and precision, the failure or disruption tolerance of the underlying service/application, the desired level of redundancy, and the likelihood of GPS disruption and potential impact. As a result, while backup solutions and processes are universally implemented within the industry, specific implementations vary widely both within a particular industry segment, and across industry segments.

Because automatic backup capabilities and other safeguarding/mitigation strategies are widely available and implemented, short-term loss or disruption of GPS will have minimal impact on the commercial communications infrastructure and its operations. One important exception is that short-term loss or disruption of GPS signals will affect the ability to determine accuratelocation information for wireless E911 purposes.

The specific consequences of medium- to long-term loss or disruption of GPS will vary based on a number of factors, including the specific function or application being supported by GPS, the duration of the loss/disruption, the geographic size of the affected region, and the availability and implementation of effective backup capabilities and contingency plans. Feedback from the study generally indicates that the wireline network infrastructure, including wireline components of wireless, satellite, cable, and broadcast networks, will sustain operation automatically for approximately 30 days. Network performance would be closely monitored, as it is still possible for performance to be impacted during this time period. For other components, the impact of long-term GPS loss varies. For example, in the wireless network environment, the ability to hand calls off between code division multiple access-based cell sites will begin to be affected after 24 hours. In the satellite, cable, and broadcast network environments, service-specific impacts unique to those environments could occur (e.g., experiencing delay in the time to acquire satellite lock, reverting to the manual recording of radio frequency signal leakage by cable network operators, experiencing In Band On Channel/Hybrid Definition radio transmission degradation in the broadcast environment).

In the extremely unlikely event of a complete and catastrophic loss of GPS over an extended period of time (e.g., more than one month) and affecting a large geographic area(e.g., nationwide, continental, global), overall impact is more difficult to ascertain. Because of the diverse and highly distributed implementations of GPS-based solutions across the industry, any impact likely would be experienced in the form of a gradual degradation of network performance, with little potential for cascading network failures. Additional backup capabilities, processes, and mitigation approaches can and will be used to sustain network operation beyond this period; however, mitigation of an extended and complete loss of GPS would require costly reconfiguration of the network to redistribute alternative timing sources. Such a reconfiguration would require a cooperative effort between carriers.

The NSTAC also emphasizes that commercial communications networks do not operate in a vacuum, and service providers and network operators will take immediate corrective actions in response to any size event, particularly a large-scale catastrophic event with the potential to degrade the network. Even before all automatic means of backup are exhausted, companies will have already executed contingency plans and performed manual reconfigurations and network timing adjustments as required to maintain network operation.

Overall, industry members surveyed believe that their companies have taken measures to safeguard against those disruptions to the GPS signal that are likely to be encountered; however, to date, no industry or Government exercise has sought to replicate the impact of a long-term or permanent GPS outage simultaneously on all industries. The NSTAC recommends that the

President direct the Department of Homeland Security and the Department of Defense to include various GPS outage scenarios in future planned disaster recovery exercises in coordination with the commercial communications industry.

1.0. BACKGROUND AND PURPOSE

In response to a January 2003 request from the Director, National Security Space Architect, the President's National Security Telecommunications Advisory Committee (NSTAC) reviewed and assessed policies, practices, and procedures for the application of infrastructure protection measures to commercial satellite communications systems used for national security and emergency preparedness communications. Specifically, the NSTAC reviewed applicable documentation addressing vulnerabilities in the commercial satellite infrastructure and identified potential policy changes that would bring the infrastructure into conformance with a standard for mitigating those vulnerabilities. As a part of its review, the NSTAC also considered Global Positioning System (GPS) timing capabilities and developed initial findings and a recommendation for further study of GPS-related issues. The results of this effort were published in the *NSTAC Satellite Task Force Report*, March 2004.

At the 2007 NSTAC Meeting, Ms. Frances Fragos Townsend, Assistant to the President for Homeland Security and Counterterrorism, requested that the NSTAC begin a scoping effort to further evaluate the commercial communications infrastructure's reliance on GPS.

Ms. Townsend called for the NSTAC to present its findings and recommendations for White House evaluation.

In response to this request, the NSTAC formed a working group comprised of industry and Government representatives to review findings from the March 2004 study and examine the commercial communications reliance on GPS, as well as the possible impacts that loss or disruption of GPS could have on the commercial communications industry, including its reliance on GPS for synchronizing local timing clocks. This response presents the NSTAC's findings for White House review and consideration.

2.0. Approach and Scope

The study approach and scope for this effort are briefly discussed below.

2.1. Approach

Representatives of NSTAC member companies, subject matter experts (SME) from non-NSTAC commercial communications companies, trade associations, Government participants, and GPS technical experts contributed to this effort. To gain a broad understanding of the use of and reliance on GPS within the commercial communications industry, the NSTAC invited SMEs from the Government, private sector, and academia to present briefings. The NSTAC also reviewed previous studies, including the March 2004 *NSTAC Satellite Task Force Report* findings on GPS vulnerabilities in the commercial satellite infrastructure and the findings and recommendations of the August 2001 *Vulnerability Assessment of the Transportation Infrastructure Relying on the Global Positioning System*, prepared by the Volpe National Transportation Systems Center. Appendix A provides a list of working group members, Government personnel, and other participants.

To gain current perspectives on the industry-wide use of GPS, the NSTAC solicited information from its members as well as representatives of the Department of Homeland Security (DHS) National Coordinating Center-administered Telecommunications Information Sharing and Analysis Center. Specifically, the NSTAC requested information on: (1) company and industry segment use of and reliance on GPS timing and precision location signals; (2) impacts to networks and operations that would result from loss or degradation of GPS signals; and(3) specific mitigation strategies implemented or planned to minimize the impact of any GPS signal loss or degradation.[1]

A broad cross-section of the commercial communications industry submitted feedback, including responses from individual companies in the telecommunications, computer software/services, and aerospace and defense industry sectors, and from industry trade associations. The NSTAC presents its findings in this document and looks forward to continued engagement with the White House staff to support ongoing national security telecommunications policy development and program planning.

2.2. Scope

This study focuses narrowly on the commercial communications industry's use of and reliance on GPS, in accordance with White House direction to tailor the effort to enable a quick response. Deliberations with SMEs and evaluation of previous studies included information involving GPS uses and applications to military, maritime, aviation, and other civil environments. While instructive in understanding overall GPS deployment trends and vulnerabilities, non-commercial information was not evaluated or integrated into the NSTAC's findings, as that area was deemed to be outside the scope of this effort.

In soliciting information and perspectives from commercial communications industry representatives and the larger SME community, the NSTAC specifically requested that data submitted for the analysis be non-proprietary and unclassified. The findings documented in this response are generally applicable across all industry segments and represent NSTAC member company consensus. However, it is important to note that the study analysis revealed significant variance in the use of and reliance on GPS across industry segments and across companies within each industry segment. Feedback from individual companies likewise indicated that a wide variety of strategies, techniques, and implementation approaches are applied to mitigate the impacts of GPS loss or disruption, reflecting company-specific business case and risk assessment determinations.

3.0. Commercial Communications Reliance on GPS

This section describes the commercial communications reliance on GPS. Section 3.1 discusses the use of and reliance on GPS, including applications and dependencies specific to the wireline, wireless, satellite, cable, broadcast, and corporate/enterprise network environments. Section 3.2 generally characterizes the impact of loss or disruption of GPS for each network environment. Section 3.3 identifies associated strategies, employed or planned by industry, to mitigate GPS-related impacts.

3.1. Use of GPS

GPS is a U.S. Government-owned utility that provides users with positioning, navigation, and timing (PNT) services. The U.S. Air Force operates the space and control segments, consisting respectively of the GPS satellite constellation and the worldwide control stations that maintain the satellite orbits and adjust the satellite clocks. The user employs GPS receiver equipment to receive signals from the satellites and calculate the user's location. Using the signals to measure the distances to at least four satellites simultaneously, a GPS receiver can determine three-dimensional position (latitude, longitude, and altitude) while synchronizing its clock with the GPS precise time standard.[2] The GPS constellation, illustrated in Figure 1, consists of a minimum of 24 satellites in one of six medium-earth orbits, approximately 20,000 kilometers above the earth's surface.[3]

The U.S. Department of Defense (DOD) began development in the 1970s of what would become the GPS system. However, the first U.S. pronouncement regarding civil use of GPS came in 1983 following the downing of Korean Airlines Flight 007. The Soviet Union shot down the airplane after it strayed over Soviet territory; afterwards, President Reagan announced that GPS would be made available for international civil use once the system became operational.

The first major success of GPS came in 1990-1991, during Operation Desert Storm. DOD's needs during the crisis sparked a surge in the GPS market, which had barely existed just a few years prior to the war. Desert Storm provided a showcase for all the military uses of GPS—from helping soldiers navigate across the desert to vastly improving targeting capabilities of artillery and bomber units. Following the war, GPS device sales to non-DOD customers surged, and U.S. commercial GPS manufacturers continue to produce new and cheaper receivers that are used across numerous industries and infrastructures.[4]

On December 8, 2004, the President established a new national *U.S. Space-based Positioning, Navigation, and Timing Policy* containing guidance and implementation actions for space-based PNT programs, augmentations, and activities for U.S. national and homeland security, civil, scientific, and commercial purposes.[5] In the policy, the U.S. Government pledged to provide on a continuous, worldwide basis, civil space-based PNT services free of direct user fees for civil,commercial, and scientific uses. The policy also established an Executive Committee that is charged in part with ensuring that efforts to deny hostile use of any space-based PNT services will not unduly disrupt civil and commercial access to civil PNT services outside an area of military operations, or for homeland security purposes.

Although DOD controls and maintains the GPS, and makes the service available to U.S. and allied armed forces, there is also a large civilian component in the user community. The Department of Transportation is responsible for overseeing all civil uses of GPS, which has become integral to navigation for aviation, ground, and maritime operations. Emergency responders depend upon GPS for location and timing capabilities in their life-saving missions. Banking, mobile phone operations, and the control of power grids are facilitated by the accurate timing provided by GPS. Farmers, surveyors, and geologists use the free and open GPS signals to pinpoint locations.[7]

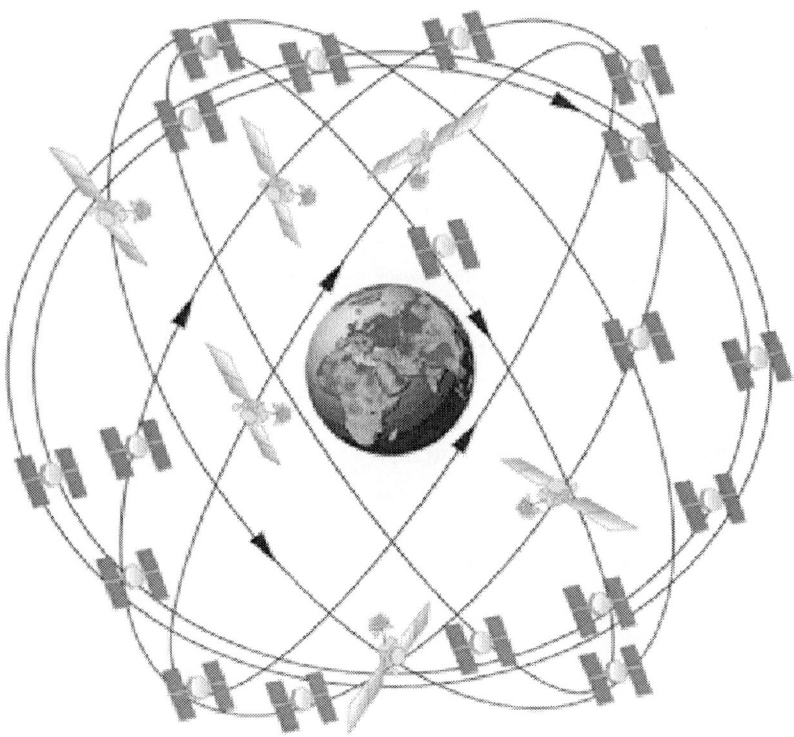

Figure 1. GPS Constellation[6].

Commercial communications companies have selected and widely implemented GPS-based network timing and synchronization solutions primarily because GPS provides an inexpensive, globally-available, highly reliable, and extremely accurate reference timing source. The U.S. Government's commitment to provide and maintain civil space-based PNT services, such as GPS, has also encouraged rapid adoption of GPS throughout the commercial communications industry. The use of GPS-disciplined oscillators (GPSDO), implemented extensively throughoutthe industry, is a cost-effective solution able to meet the various stringent network performance requirements for time and frequency. The use of and reliance on GPS in the wireline, wireless, satellite, cable, broadcast, and enterprise network environments are further described below.

Wireline Network Environment

GPS signals in the wireline network environment are fundamentally used as a primary reference timing source for a diverse range of telecommunications network equipment, including wireline switching offices, Synchronous Optical NETwork (SONET) and Synchronous Digital Hierarchy (SDH) nodes, multiplexer and demultiplexer equipment, digital cross connects, and customer premise equipment (e.g., private branch exchanges [PBX]). GPS is also used as a reference timing source for other industry segment wireline-connected network elements, including mobile switching centers (MSC), satellite network control and earth station equipment, and other core cable network and broadcast network elements.

In addition to provision of a primary time reference, GPS signals also are used to support essential network time, frequency, and phase synchronization functions. Digital telecommunications networks require highly reliable precision frequency and timing information to maintain data integrity and guarantee the delivery of high quality services. Timing impairments, or "slips," can cause impacts to service quality such as increased noise or "pops" during voice calls, loss of picture content during facsimile transmission, inefficient retransmission of data packets, and video content "drop out" and "freeze frame" occurrences. More severe impacts resulting from timing and synchronization impairments include dropped calls/connections, the inability to initiate/receive calls and establish/maintain connections, the loss of circuit or transmission path integrity, and eventual network element isolation and/or placement in an "out of service" condition. As the commercial communications network infrastructure continues to evolve toward a high-speed all-digital environment, accurate timing and synchronization functions that support the infrastructure are becoming more critical.[8]

In characterizing their use of and reliance on GPS, companies across industry segments(e.g., wireless, satellite, cable, and broadcast) note the associated wireline network reliance on GPS as a potential factor in their own reliance on GPS. Network interconnections (e.g., network interface points) and leased lines used to connect internal network elements are examples of underlying wireline components whose operation may have a dependence on GPS.

Wireless Network Environment

GPS is used to provide the highly accurate timing source required to synchronize mobile phones to the cellular network and to synchronize cellular network elements to one another. Radio carrier frequencies must also be synchronized precisely in order to prevent co-channel interference (i.e., cross talk) and other radio frequency (RF) interference problems. For code division multiple access- (CDMA) based networks in particular, precise frequency synchronization is required to support handoff of calls between cell sites.

Another important use of GPS in the cellular network environment is in support of location-based services, including support of Enhanced 911 (E911) Phase II requirements. To comply with the Federal Communications Commission's (FCC) wireless E911 requirements,wireless carriers require accurate positioning information to provide the precise locations of wireless network callers so that police, fire, and emergency rescue personnel can be dispatched quickly.[9]

GPS also provides Time-of-Day information to support several cellular network functions including setting clock time on mobile devices and accurately time stamping billing records and data packets for network and service performance measurement. Feedback from industry also identified internal company use of wireless network services (e.g., use of cellular and paging services by company employees and contractors) as a potential peripheral reliance on GPS. Internal communications may be disrupted if these wireless network services become unavailable due to loss of GPS.

Satellite Network Environment

In the satellite industry, use of GPS signals is fundamental in providing timing reference and synchronization across satellite constellations and satellite network elements. In addition to timing synchronization, satellite network operators use GPS signals broadly in support of telemetry, tracking, and control (TT&C) time tracking and ranging operations as well as frequency referencing for many applications. Most land earthstations (LES) use GPS as a primary means to set the internal station frequency standard and clocks, and central satellite control systems use GPS as a master clock timing reference.

Fixed satellite-based communications terminals also use GPS for geo-location and a timing reference. GPS capabilities enable terminals to quickly locate and acquire a satellite. Terminals may also use GPS to synchronize the terrestrial communications equipment with which they interface, providing consistent data flow throughout the link. Some satellite terminals that use complex spread spectrum waveforms rely heavily on GPS for system synchronization.

Cable Network Environment

The cable television industry also relies on GPS time signals as a primary timing reference for several network infrastructure components, including Building Integrated Timing Supply (BITS) clocks, Network Time Protocol (NTP) server deployments, Data Over Cable Service Interface Specification (DOCSIS) timing interface servers, and interconnecting or supporting wireline and wireless network elements (e.g., time-division multiplexing [TDM] circuits, T1 emulation circuits, microwave radio links). NTP servers are used by cable operators in support of the following: synchronization across cable system equipment such as servers, routers, switches, and terminal equipment (e.g., cable modems); time stamps in simple network management protocol tables and local logging for error and event correlation; set-top boxes, which use time information for use in electronic program guides, reliable content recording, and in processing Emergency Alert messages; and service applications such as video-on-demand (VOD) and Ad Insertion systems. Additionally, pursuant to FCC regulations, cable operators routinely monitor their systems for RF signal egress or "leakage," using signal leakage detection equipment that utilizes GPS location signals to automate the process of locating and repairing signal leaks.

Broadcast Network Environment

GPS-based systems are widely employed by broadcast radio and television (TV) stations. GPS provides precise timing and phasing references for equipment throughout the broadcast production and transmission chain. GPS is used as frequency reference for both analog and digital television transmitters. This is particularly important in the proper implementation of "Precision Off-Set," which is used to ensure that a digital TV (DTV) transmitter will not interfere with an analog transmitter operating on the same channel. In addition, single frequency TV networks recently have been approved for use by the FCC and are just beginning to be deployed in the U.S. This technology, called Distributed Transmission, relies heavily on GPS to ensure that all the transmitters in the network remain synchronized.

GPS is used to support the proper synchronization of digital radio transmitters. The U.S. In-Band-On-Channel (IBOC) digital transmission technology, also known as Hybrid Digital (HD) radio, overlays digital carriers onto an FM station's analog signal and relies on precise timing to ensure that the digital signal does not degrade those analog transmissions.

In both radio and TV production studios, GPS is used as the reference for the master clock system, which ensures that all the clocks in the studio increment their second hands simultaneously and remain locked to the same time. GPS is also used to derive master timing reference signals, which are required in a broadcast production system to keep audio and video in synchronization, to ensure that automation systems are time-aligned, and to meet the various requirements for broadcast standards compliance.

Enterprise Network and Corporate Operations Environment

Feedback from the commercial communications industry also identifies use of and reliance on GPS signals to support company-internal enterprise network operation as well as corporate operation functions. For example, one company cites use of GPS timing via NTP servers for synchronization of all device timing (e.g., timing for switches, routers, servers, and desktops) on its global corporate network. Multiple GPS receivers feed timing data to distributed NTP servers, which synchronize their clocks to the GPS-provided time reference on a periodic basis. Network and computing resources then access the NTP servers for timing. The company also cites use of GPS timing to synchronize its internal SONET infrastructure. Regarding corporate operations, several companies cite extensive use of communications services (e.g., cellular and paging services, satellite phone service, use of personal communications devices such as Blackberry® devices) by employees and/or contractors while performing their jobs. Availability of these communications services may be impacted by a GPS loss or disruption. GPS equipment vendors also utilize the GPS signal during product development, testing, and production.

In support of workforce and resource management functions, GPS signals are used across the industry by field operations staff to more effectively coordinate service and maintenance activities. For example, in support of the cable industry's field service, plant maintenance, and auditing activities, company vehicles may utilize GPS signals as a part of intelligent automated vehicle fleet management systems. Another cited example of use of GPS positioning data is a disaster recovery and employee location mapping capability used by the company in the event of a regional incident. GPS location information is also used by some companies to support industry functions such as fiber locating operations (or "call before you dig" operations) and other field test and measurement functions. As the overall market for GPS-based devices and services continues to grow, the commercial communications industry is likely to identify andutilize additional uses of GPS to increase productivity, service delivery, and the number of available end-user applications. Table 1lists some examples of the commercial communications industry's uses of and reliance on GPS signals.

Key findings regarding the commercial communications industry's use of and reliance on GPS are:

- The U.S. Government's commitment to provide and maintain civil space-based PNT services, such as GPS, free of direct user fees for civil, commercial, and scientific uses has encouraged rapid adoption of GPS throughout the commercial communications industry.
- GPS supports a broad range of commercial communications industry functions and applications in many commercial communications industry segments (e.g., wireline, wireless, satellite, cable, and broadcast network environments).
- The primary use of GPS in the commercial communications industry and across all commercial communications industry segments is the support of precision timing and network synchronization functions.
- Another important use of GPS signals is support to location-based services, including support of wireless E911 Phase II requirements.
- As the commercial communications network infrastructure continues to evolve toward a high-speed all-digital environment, accurate timing and synchronization functions that support the infrastructure are becoming more critical.
- As the overall market for GPS-based devices and services continues to grow, the commercial communications industry is likely to identify and utilize additional uses of GPS to increase productivity, service delivery, and the number of available end-user applications.

3.2. Impact of Loss or Disruption of GPS

In its 2004 study, the NSTAC found that "impacts of a loss of GPS could be seen across all aspects of the telecommunications industry ... on wireline and wireless networks ...[and] fiber optic and broadband transmission systems; radio, television, and cable broadcast systems; and satellite systems – all of which use GPS to some extent for synchronizing local timing clocks." Industry submissions in this GPS study generally confirm this finding and further substantiate specific impacts of GPS loss or disruption within each industry segment.

Generally, feedback indicates that short-term loss or disruption of the GPS signals for timing will have minimal impact on the commercial communications infrastructure and its operations. One important exception is that short-term loss or disruption of GPS signals will affect the ability to determine accurate location information for wireless E911 purposes. The impact of medium- to long-term loss or disruption of GPS will vary based on a number of factors, including thespecific function or application being supported by GPS, the duration of the loss/disruption, the geographic size of the affected region, and the availability and implementation of effective backup capabilities and contingency plans.

Specific impacts of GPS loss in the wireline, wireless, satellite, cable, broadcast, and enterprise network environments are described below. Additional details on strategies for mitigating the impacts of GPS loss are presented in Section 3.3.

Wireline Network Environment

In the case of a short-term complete GPS loss or a long-term localized GPS loss, carriers indicate that the impact on wireline network operation is minimal due to the availability and use of backup systems and processes, alternative timing sources, and effective business continuity planning.[10] The most commonly-cited potential impact in the wireline network environment is the eventual loss of network timing and synchronization as a result of a long-term complete loss or disruption of the GPS timing signal across an extended area. Wireline carrier feedback indicates that wireline network infrastructure (e.g., circuit switches) will sustain operation automatically for approximately 30 days. Network performance would be closely monitored, as it is still possible for performance to be impacted during this 30-day window. Secondary and tertiary backup capabilities and other mitigation processes can and will be used to sustain network operation beyond this period. Carriers also note that mitigation of an extended and complete loss of GPS beyond this period would require costly reconfiguration of the network to redistribute alternative timing sources. Such a reconfiguration would require a cooperative effort between carriers.

Table 1. Examples of GPS Use and Reliance

Network Environment	Application/Use
Wireline	• Time/frequency reference source (e.g., central offices [CO], SONET, TDM circuits, digital access and cross-connect (DAC) systems, termination equipment, voice switches) • Network timing/synchronization • Workforce and Resource Management
Wireless	• Time/frequency reference source (e.g., wireline elements, MSCs, cell sites, HLR/VLRs, mobile devices) • Network timing/synchronization (network element-to-network element, mobile phone-to network, RF carrier frequency sync) • CDMA mobile unit handoff • E911 Phase II and location-based services • Time-of-Day functions (clocks on mobile units, timestamp for billing and performance measurement) • Workforce and Resource Management
Satellite	• Time/frequency reference source (e.g., satellite network ground segment elements, land earth stations, TDM circuits, satellite terminals) • Network/Application timing and synchronization • TT&C time tracking and ranging operations • Workforce and Resource Management
Cable	• Time/frequency reference source (e.g., switches, routers, NTP server deployments, DOCSIS timing interface servers, cable modems, set-top boxes) • Network timing and synchronization • Set-top box use (electronic program guides, content recording, alert message processing) • Service applications (VOD, Ad Insertion systems) • RF signal leakage detection • Workforce and Resource Management
Broadcast	• Time/frequency reference source (e.g., broadcast production/distribution facilities, digital TV "Precision Off-Set") • Network timing and synchronization (e.g., DTV, IBOC/HD radio) • Broadcast audio/video synchronization • Time alignment for Automation systems • Per occasion Applications (e.g., remote broadcasting) • Workforce and Resource Management
Enterprise/ Corporate Operations	• Timing/frequency reference source • Network and device timing and synchronization (SONET, servers, routers, switches, desktops) • Fiber locating • Workforce and resource management • Development and production of GPS equipment and devices

It should be noted that such an event, resulting in complete loss of GPS for an extended time and over a large geographic area, has never occurred.[11]Additionally, no industry or Government exercises have sought to replicate the impact of a long-term or permanent GPS outage simultaneously on all industries.

CO timing signal generator (TSG) systems provide a common source for frequency and phase alignment of all network elements operating in the CO building. This synchronization is essential for interoperability of digital transmission networks. The TSG receives timing from a highly accurate GPS primary reference source (PRS), cesium PRS, or Stratum 1-traceable timing delivered via an interoffice facility. These timing systems provide a robust, simple-toadminister, and trouble-free network of clocks of known quality and performance characteristics. Reliable clocks ensure that network synchronization provides the necessary level of performance demanded by a growing digital network.

The hierarchy of clock requirements is grouped into four stratum levels, as defined by the American National Standards Institute (ANSI) T1.101 standard.[12] The standard defines the minimum performance requirements for telecommunications network synchronization and timing requirements for each stratum level, as shown in Table 2.

- Stratum 1 is the highest quality level in the clock hierarchy. Stratum 1 clocks are defined as autonomous sources, requiring no input from another source. In order to meet interface standards, all digital signals must be under the control of a clock or clocks traceable to a Stratum 1 source. Stratum 1-level timing sources, typically atomic oscillators (e.g., cesium beam) or GPSDOs, are specified to have a maximum "drift" of 1×10^{-11}. As shown in Table 2, T1 carrier cycle slips can be expected to occur only once every 72.3 days, worst case, if Stratum 1-quality clocks are used.
- Stratum 2 and lower-level clocks require input and adjustment from a higher stratum-level clock. Stratum 2 clocks are typically used as the master TSG oscillator at critical network sites. Stratum 2 TSG systems employ rubidium oscillators for extended holdover capability.
- Stratum 3E clocks are used as the master TSG oscillator at other locations in the network that are not Stratum 2 equipped. The Stratum 3E level was defined as a result of the widespread deployment of SONET transport and the associated need for enhanced phase filtering capabilities. Stratum 3 clocks are used in digital switches, DACs, and SONET network elements.

- Stratum 4 clocks are found in distribution facilities (e.g., channel banks) and end-user switching equipment (e.g., PBX).

It should be noted that every network element and every clock is effectively operating at the Stratum 1 level when the timing hierarchy is intact. The stratum level of subtending clocks only becomes a factor when the timing distribution chain is disrupted, and the holdover characteristics of the oscillators come into play.

Table 2. Stratum Clock Hierarchy and Timing Accuracy Requirements[13]

Stratum Levels	Stratum 1	Stratum 2	Stratum 3E	Stratum 3
Frequency accuracy, adjustment range	1×10^{-11}	1.6×10^{-8}	1×10^{-6}	4.6×10^{-6}
Frequency stability	NA	1×10^{-10}	1×10^{-8}	3.7×10^{-7}
Pull-in range	NA	1.6×10^{-8}	4.6×10^{-6}	4.6×10^{-6}
Time offset per day due to frequency instability	0.864 μs	8.64 μs	864 μs	32 ms
Interval between cycle slips	72.3 days	7.2 days	104 minutes	169 seconds

In the event of a complete loss of GPS signals, the wireline synchronization network is designed to fall back on internal network clocks, such as cesium PRS systems and rubidium and crystal oscillators, used for extended holdover capability. As Stratum 1-level clocks, cesium PRSsystems are autonomous timing sources, equal in quality to a GPS-derived timing signal. Due to cost considerations, cesium PRS systems generally are only deployed at critical network sites.

For locations that receive their timing reference from a GPS PRS, extended loss of GPS would eventually cause the oscillator in the TSG to enter holdover status. Once in holdover, the oscillator in the TSG can maintain accurate frequency timing for a period dependent on the type of oscillator. For a Stratum 2 rubidium oscillator, network performance will be maintained for about thirty days. For a Stratum 3E crystal oscillator, network performance will be maintained for seven to ten days.[14] Once the holdover capability of the TSG oscillator is exceeded, these clocks would begin to "drift" away from a common frequency, and network elements would gradually lose synchronization with one another.

Service providers also cite the use of available external timing reference sources as a means to establish an accurate time reference (e.g., geographically diverse and redundant GPS-based devices, a backup precision timing reference source such as the LORAN-C signal, reconfiguration to “line time” off an interconnected network). Approaches to timing and synchronization backup (e.g., the types and order of secondary and tertiary backup sources employed) vary by service provider; however, all major carriers adhere to Telcordia standards for timing synchronization.[15]

As noted in the 2004 NSTAC study, a general approach in the public switched telephone network (PSTN) is to deploy a Stratum 1 timing source to every CO through a combination of cesium PRS systems, GPS-based solutions, and interoffice distribution of Stratum 1-traceable timing references.[16] Cost remains a primary factor in selecting a solution. For example, the cost of a cesium-based solution typically exceeds that of a GPS-based solution by about $20,000. Interoffice distribution of timing references is the least capital-intensive solution, but requires extensive planning and maintenance to ensure proper execution.

In summarizing impact to the wireline network, carrier feedback indicates that the wireline network infrastructure (e.g., circuit switches) will sustain operation automatically for approximately 30 days in the event of complete loss of GPS signals. Network performance would be closely monitored, as it is still possible for performance to be impacted during the 30-day window. Additional backup capabilities, processes, and mitigation approaches can and will be used to sustain network operation beyond this period. However, mitigation of an extended and complete loss of GPS would require costly reconfiguration of the network to redistribute alternative timing sources. Such a reconfiguration would require a cooperative effort between carriers.

Wireless Network Environment

Long-term GPS timing signal loss or degradation could impair wireless network timing and synchronization. GPS is utilized in wireless network synchronization and provides a precise timing and frequency reference source for cell site radio controllers, MSCs (and interconnected wireline switching offices), and other wireless network elements. All of these network elements have backup internal and/or external timing sources, but if the GPS clock source is lost or disrupted, the internal timing sources will begin to drift from component synchronization at a rate based upon the class/type of clock implemented, and the timing error will build proportionally over time during which the reference source is unavailable. The error rate will build over time, and hard failures will manifest themselves randomly once the clocks drift outside of the system synchronization thresholds.

Wireless network operators report that the cell sites in an affected area likely would be the first cellular network elements to begin to drift, as their internal clocks typically guarantee only24 hours of highly accurate holdover time. Beginning after 24 hours and as the timing of the cell sites drifts apart from one another (due to the lack of a common time reference), handoffs between cell sites would begin to fail, and cell sites would start to become isolated from the other cell sites in the network; however, in this scenario, cell sites would still be able to communicate with the MSC, and calls could still be originated from subscriber phones.

The wireline components of cellular networks can be expected to perform as discussed in the wireline network environment section above. For example, the lack of synchronization between MSCs would begin to result in "slips" on digital inter-office circuits/elements, eventually affecting circuit integrity between MSC locations and resulting in communication issues and data loss between network offices. It is likely that cellular telephone customers would initially experience temporary minor communication issues (e.g., pops, clicks, and data loss), which would worsen until the connection to the cell site was effectively out of service.

One wireless service provider noted that even though the company's time reference is maintained with multiple high quality reference time sources, communication with network elements external to the company would also depend upon the ability of those external elements to maintain an accurate time reference. Another wireless service provider stated that, in the event of long-term loss or disruption of GPS, its wireline portion of the network would likely remain operational indefinitely due to redundant backup capabilities (i.e., externalStratum 1-quality timing source as a primary backup and rubidium-based oscillators as a secondary backup).

In addition to timing and data synchronization impacts, loss or degradation of GPS-based positioning information would critically impact wireless Phase II E911 and commercial location-based services. These services would immediately suffer from the inability to gather precise ranging measurements from satellites currently in the visible horizon during a GPS outage. Having fewer operational satellites in the GPS constellation or being out of the range of the receiving site would result in fewer possible location measurement points, making the determination of a highly accurate position estimate more difficult or impossible.[17]Strategies to mitigate the impact of loss of GPS-based location data are further discussed in Section 3.3.

Other impacts of GPS loss in the wireless network environment include loss of Time-of-Day data that could affect billing and measurement systems' accuracy.

In summarizing the impact on wireless networks, carrier feedback indicates that the first network elements likely to be affected are cell sites, which will sustain operation for at least 24 hours. After this time period, the ability to hand calls off between cell sites will be affected, although calls can still be originated from subscriber phones as communications with the MSC will not be affected. Wireline network elements generally will sustain operation automatically for up to30 days, although wireless carrier estimates of this time period varied from five days to beyond 30 days.

Satellite Network Environment

The 2004 NSTAC study noted that "most satellite operators use GPS timing for TT&C time tracking, ranging operations, and timing synchronization." Current study responses concur with previous the NSTAC findings. In the event of complete loss of GPS, one satellite network operator notes that its GPS receiver equipment is able to operate independently for several days. After this time period, available backup cesium-standard clocks and stable clock generators would be used as input timing reference to the GPS receiver equipment. This approach offers a long-term solution until GPS-based satellite timing is restored. The satellite network operator notes that it also is investigating the potential for its equipment to accept an external Inter-Range Instrumentation Group (IRIG-H) signal to use for synchronization via the National Institute of Standards and Technology (NIST) WWVB signal.[18]

Another satellite network operator notes that the absence of GPS does not cause an immediate threat to commercial satellite fleets' health; however, the capability to monitor and control satellite fleets would degrade gradually. Backup timing synchronization can be obtained from other sources such as NTP servers, but those servers may be dependent on GPS signals. Manual synchronization is also possible; however, it may prove unsustainable in the long term. One satellite service provider stated that failure of GPS "would be an inconvenience to the satellite control system" and would not result in loss of control.

LESs and satellite terminals are other satellite network elements whose operation may be impacted by GPS disruption. Most LESs use GPS to set the internal station frequency standard and clocks. They typically have atomic clocks for backup timing; however, one respondent noted that localized GPS anomalies may have to be resolved prior to backup initiation, resulting in a temporary outage. Some satellite user terminals require a GPS signal for location, spot beam designation and timing, while other user terminals may have access to platform navigation systems for timing. Other terminals do not require GPS or any external navigation system to function.

Another company response noted that satellite terminal designs are becoming increasingly dependent on GPS capabilities. When satellite terminals employ GPS receivers for geo-location, loss of GPS would impact the ability to quickly find, acquire, and track a satellite. Some fixed satellite-based communications terminals routinely use GPS to increase signal acquisition speed; a loss of the GPS signal would lengthen the acquisition time for an affected terminal.

The potential impact on fixed satellite-based communications terminal operation also can vary depending on terminal design and the underlying technology employed. For example, a simple frequency division multiple access terminal designed with a rubidium-based backup solution could operate almost indefinitely without GPS timing. However, satellite terminals that use more complex spread spectrum waveforms may be more dependent on accurate timing. A spread spectrum application that is not protected by an atomic frequency standard backup solution may suffer acquisition time degradation after only a few hours of GPS signal loss.

In summarizing feedback regarding the satellite network environment, network operators indicate that loss of GPS has minimal impact in the short term. Impacts of a long-term complete GPS loss will vary by company; backup capabilities and processes are available and will be used to mitigate potential impacts.

Cable Network Environment

The cable industry's wireline network infrastructure would be subject to the same wireline-associated impacts of GPS loss or disruption as previously described. The lack of GPS time signals for a prolonged time period would result in frame slips for TDM circuits, T1 emulation, and SONET systems, which would eventually impact the ability for these circuits and networks to carry traffic without some degradation. The lack of GPS time signals would also potentially result in inaccurate clocks on NTP servers used for synchronization across network equipment, set-top boxes, cable modems, and service applications. Potential impairments include inaccurate electronic program guide data, incorrect billing of digital voice calls, and prolonged debugging of network errors due to the lack of synchronized clocks. Should cellular communications become unavailable due to loss or disruption of GPS, cable operations (e.g., work force coordination and management) could be significantly impacted. The use of GPS signals in support of vehicle fleet management is in a nascent stage within the cable industry and impact on productively is likely minimal at this time; however, the loss of productivity may be greater in the future when cable operators have near real-time location information integrated into automated vehicle routing and dispatch systems.

Another associated cable industry impact of GPS loss or disruption is the inability of RF signal leakage detection equipment to automate the process of precisely locating the signal leak in a cable system. Manual recording of leakage locations can be used; however, that method is likely to be less accurate than the automated methods that utilize the GPS location signals, and may lengthen the time needed to repair the signal leak.

In summarizing feedback regarding the cable network environment, the cable network infrastructure dependent upon GPS time signals will sustain operation automatically for approximately 30 days in the event of complete loss of GPS signals. Beyond this period, some circuit termination equipment could be reconfigured to utilize the receive clock from the PSTN as a reference clock; however, this option may not be viable in the event of widespread GPS outage or degradation.[19]

Broadcast Network Environment

GPS is not critical to the operation of most broadcast systems. In the studio, most equipment components, including the master clock and timing referencesignal generators, have their own internal oscillators, which are very stable. If GPS fails, this equipment can be set to "manual" or will automatically revert to the internal oscillators and will be able to operate for some time without drifting off frequency. The same is fundamentally true for digital and analog television transmitters. While they rely on GPS for synchronization and frequency reference, they also have very stable internal oscillators that will take over in the event that GPS fails.

The areas for which GPS is critical are: (1) "Precision Off-Set" between analog and digital television transmitters (this area will no longer be critical after February 2009 when full service analog TV transmitters permanently stop broadcasting as required by law); (2) IBOC transmitters, which may interfere with companion analog FM signals; and (3) Distributed Transmission networks for television, in which the loss of a full-time frequency and time reference at the transmission sites will result in the transmitters creating interference with each other.

In general, through use of internal reference sources, these systems can "flywheel" through a loss of GPS synchronization for an extended period of time.[20]The severity and frequency of intermittent failures of the transmitted signals is directly proportional to the precision tolerance of the internal references. Eventually, system failure can occur due to a loss of synchronization; however, the amount of time to system failure is not easily predicted as it is dependent on the stability of the oscillators in each part of the overall transmission chain.

Enterprise Network and Corporate Operations Environment

GPS disruption may impact the commercial communications industry's enterprise network operation functions through loss of enterprise network timing synchronization. For example, one responder noted that, in the event of an extended GPS disruption or failure, the NTP servers, which provide timing for switches, routers, servers, and desktops, would eventually experience a time shift from true time. Long-term results could include network and service outages, discrepancies in security logs, invalidated public key infrastructure certificates and tokens, and SONET infrastructure failure. Corporate operations functions may also be impacted as industry employees and contractors lose the ability to communicate via paging, cellular, and personal communications (e.g., Blackberry® devices) services impacted by GPS disruption.

GPS disruption may also impact equipment vendor corporate operations with the loss of capability to use GPS broadcast signals while developing and manufacturing company products. This would interfere with product development efforts and subsequently increase development costs, delay product manufacturing and delivery rates, and increase the risk to product operational reliability. The loss or degradation of GPS broadcast signals during development would affect the ability to completely evaluate and test operational system capabilities before production.

In summary, key findings regarding the impact of loss or disruption of GPS are:

- Generally, short-term loss or disruption of the GPS signals for timing will have minimal impact on the commercial communications infrastructure and its operations.
- Short-term loss or disruption of GPS signals will affect the ability to determine accurate location information for wireless E911 purposes.

- The impact of medium- to long-term loss or disruption of GPS will vary based on a number of factors, including the specific function or application being supported by GPS, the duration of the loss/disruption, the geographic size of the affected region, and the availability and implementation of effective backup capabilities and contingency plans. For example:
  - In the event of complete loss of GPS signals, wireline carrier feedback indicates thatwireline network infrastructure (e.g., circuit switches) will sustain operation automatically for approximately 30 days. Network performance would be closely monitored, as it is still possible for performance to be impacted during the 30-day window. Additional backup capabilities, processes, and mitigation approaches can and will be used to sustain network operation beyond this period; however, mitigation of such an extended GPS loss would require costly reconfiguration of the network to redistribute alternative timing sources. Such a reconfiguration would require a cooperative effort between carriers.
  - In the event of complete loss of GPS signals, wireless carrier feedback indicates that the first network elements likely to be affected are cell sites, which will sustain operation for at least 24 hours. After this time period, the ability to hand calls off between cell sites will be affected, although communications with the MSC will not be affected. Wireline network elements generally will sustain operation automatically for up to 30 days, although wireless carrier estimates of this time period varied from five days to beyond 30 days.
  - Feedback from the satellite operators indicates that impacts of a long-term complete GPS loss will vary by company, and that backup capabilities and processes are available and will be used to mitigate potential impacts.
  - In the event of complete loss of GPS signals, feedback from the cable network operators indicates that the cable network infrastructure dependent upon GPS time signals will sustain operation automatically for approximately 30 days. Beyond this period, some circuit termination equipment could be reconfigured to utilize the receive clock from the PSTN as a reference clock; however, this option may not be viable in the event of widespread GPS outage or degradation.

- Feedback from the broadcast industry indicates that GPS is not critical to the operation of most broadcast systems. Systems that may be affected by GPS loss include "Precision Off-Set" between analog and digital TV transmitters, HD Radio, and Distributed Transmission networks for television. For these systems, loss of GPS can be tolerated for "an extended period of time," although this time period is not easily predicted and requires further study.

- In the extremely unlikely event of a complete and catastrophic loss of GPS over an extended period of time (e.g., more than one month) and affecting a large geographic area(e.g., nationwide, continental, global), overall impact is more difficult to ascertain. Such an event has never occurred, and, to date, no industry or Government exercises have sought to replicate the impact of a long-term or permanent GPS outage simultaneously on all industries.

3.3. Mitigation Strategies

The commercial communications industry employs a range of strategies to mitigate the impact of loss or disruption of GPS. In all communications network environments, backup solutions are deployed at the most critical nodes to protect against the loss of a GPS-provided timing reference source. Generally, service providers and network operators select backup solutions and associated implementation approaches that are specifically designed to meet the requirements of the service/application. The selection of an alternative is also an economic and business case decision that must factor in available equipment types and cost, the required level of quality and precision, the failure or disruption tolerance of the underlying service/application, the desired level of redundancy, and the likelihood of impact. As a result, implementation of backup solutions vary widely both within a particular industry segment, and across industry segments. For example, at critical nodes (e.g., wireline COs, mobile switching centers, satellite control centers), redundant Stratum 1-level sources are often deployed and further protected by secondary and/or tertiary sources. For less critical applications (e.g., the time of day on adesktop computer, some NTP server applications), less accurate timing sources may be sufficient (e.g., internal quartz oscillators, internal central processing unit clocks in devices).

To protect commercial communications network elements, service providers, network operators, and vendors report a wide variety of safeguarding/mitigation approaches and contingency plans. Strategies cited to mitigate the impact of loss or disruption of GPS timing signals include:

- Use of external/internal precision cesium-based devices (e.g., beam clocks and oscillators);
- Use of external/internal rubidium-based devices;
- Use of quartz oscillators;
- Use of multiple geographically dispersed GPS receivers;
- Automatic fall-over to other timing sources in the event of primary failure (e.g., LORAN-C timing source);
- "Line timing" off other carriers' signals and "slaving" to the wireline carrier circuit blocking;
- Manual reconfiguration to receive clock timing from the PSTN;
- Manual reconfiguration to use ad hoc timing sources during an emergency;
- Use of other timing sources such as NTP servers or an IRIG source;
- Dependence on the internal free-running oscillator of the device; and
- Use of a combination of the above strategies.

Depending on the network service or application being protected, individual companies may choose to deploy primarily one type of strategy or a combination of strategies implemented in layers to provide secondary and tertiary levels of protection.

Wireless service providers cite use of terrestrial measurements as a secondary approach to locating wireless 911 callers in the event that GPS-provided location data is unavailable.[21] One service provider noted that terrestrial network measurements would be utilized until the mobile device can no longer communicate with more than one cell site. When the device is no longer able to communicate with more than one cell site, only network identification parameters and ranging measurements to the identified cell site would be utilized. Location determination solutions of this nature result are significantly less accurate than GPS-based measurements. Another wireless provider reported that no mitigation is available for E911 caller location information.

Responses from cellular service providers are also consistent with the NSTAC's previous examination of GPS and E911 geo-location. The 2004 report found that "wireless carriers typically use GPS assist technology as one means of providing the geo-location of a 911 caller. ... Different carriers use combinations of GPS and triangulation to determine the location of a wireless caller with pinpoint accuracy." The report also noted that without a properly functioning GPS system, cell site triangulation can be utilized to deliver location information, but in many areas of the country, the geographic arrangement of cell sites make the process of triangulation difficult, if not impractical. As a result, "the loss of GPS could leave 911 centers without the ability to automatically receive the location of wireless callers to 911, thereby endangering life and property. ... [This impact] would be most severe in areas with low density of cell sites, particularly rural areas and highways. Unfortunately, these are areas in which emergency rescue personnel typically most need precise location information because they must cover large areas."

In other network environments, strategies to protect against loss of GPS positioning data generally entail manual measurement and recording. For example, cable operators can fall back to the manual recording of RF signal leakage locations, which will be less accurate than automated methods that utilize GPS signals, and will lengthen the repair time for the cable leakage, resulting in less technical staff productivity.

Regarding strategies to mitigate workforce management impacts of loss or disruption of GPS, company employees and contractors would fall back to use of other existing forms of communication (e.g., two-way radios, other carriers' services, email) should there be a GPS-related loss to primary communications modes. In the area of fleet vehicle management, one response noted that no effective alternatives exist to compensate for loss of location data as aresult of loss or degradation of GPS; however, the impact on current operations is characterized as minimal. As GPS is more fully integrated into automated vehicle routing and dispatch systems, it is anticipated that companies will put in place backup processes and systems to compensate for GPS signal loss or disruption.

Key findings regarding mitigating the impact of the loss or disruption of GPS are:

- To protect critical functions such as network timing and synchronization, companies employ multiple layers of backup capabilities, mitigation strategies, and contingency plans to provide protection against GPS outages and disruptions.
- Technological, economic, and regulatory considerations necessarily factor into individual company decisions; therefore, specific mitigation strategies and backup capabilities will vary.

SUMMARY

In evaluating the commercial communications industry's use of and reliance on GPS, the NSTAC finds that:

- The U.S. Government's commitment to provide and maintain civil space-based PNT services, such as GPS, free of direct user fees for civil, commercial, and scientific uses has encouraged the rapid adoption of GPS-based solutions throughout the commercial communications industry.
- GPS supports a broad range of commercial communications industry functions and applications in many commercial communications industry segments (e.g., wireline, wireless, satellite, cable, and broadcast network environments).
- The primary use of GPS in the commercial communications industry and across all commercial communications industry segments is the support of precision timing and network synchronization functions.
- Another important use of GPS signals is support to location-based services, including support of wireless E911 Phase II requirements.
- As the commercial communications network infrastructure continues to evolve toward a high-speed all-digital environment, accurate timing and synchronization functions that support the infrastructure are becoming more critical.
- As the overall market for GPS-based devices and services continues to grow, the commercial communications industry is likely to identify and utilize additional uses of GPS to increase productivity, service delivery, and the number of available end-user applications.
- To protect critical functions such as network timing and synchronization, companies employ multiple layers of backup capabilities and other mitigation strategies to provide protection against GPS outages and disruptions.
- Technological, economic, and regulatory considerations necessarily factor into individual company decisions; therefore, specific mitigation strategies and backup capabilities will vary.
- Generally, short-term loss or disruption of the GPS signals for timing will have minimal impact on the commercial communications infrastructure and its operations.
- Short-term loss or disruption of GPS signals will affect the ability to determine accurate location information for wireless E911 purposes.

- The impact of medium- to long-term loss or disruption of GPS will vary based on a number of factors, including the specific function and application being supported byGPS, the duration of the loss/disruption, the geographic size of the region beingimpacted, and the availability and implementation of effective backup capabilities. For example:

    - In the event of complete loss of GPS signals, wireline carrier feedback indicates thatwireline network infrastructure (e.g., circuit switches) will sustain operation automatically for approximately 30 days. Network performance would be closely monitored, as it is still possible for performance to be impacted during the 30-day window. Additional backup capabilities, processes, and mitigation approaches can and will be used to sustain network operation beyond this period; however, mitigation of such an extended GPS loss would require costly reconfiguration of the network to redistribute alternative timing sources. Such a reconfiguration would require a cooperative effort between carriers.
    - In the event of complete loss of GPS signals, wireless carrier feedback indicates that the first network elements likely to be affected are cell sites which will sustain operation for at least 24 hours. After this time period, the ability to handoff calls between cell sites will be affected, although communications with the MSC will not be affected. Wireline network elements generally will sustain operation automatically for up to 30 days, although wireless carrier estimates of this time period varied from five days to beyond 30 days.
    - Feedback from the satellite operators indicates that impacts of a long-term complete GPS loss will vary by company and that backup capabilities and processes are available and will be used to mitigate potential impacts.
    - In the event of complete loss of GPS signals, feedback from the cable network operators indicates that the cable network infrastructure dependent upon GPS time signals will sustain operation automatically for approximately 30 days. Beyond this period, some circuit termination equipment could be reconfigured to utilize the receive clock from the PSTN as a reference clock;

however, this option may not be viable in the event of widespread GPS outage or degradation.

- o Feedback from the broadcast industry indicates that GPS is not critical to the operation of most broadcast systems. Systems that may be affected by GPS loss include "Precision Off-Set" between analog and digital TV transmitters, HD Radio, and Distributed Transmission networks for television. For these systems, loss of GPS can be tolerated for "an extended period of time" although this time period is not easily predicted and requires further study.

In the extremely unlikely event of a complete and catastrophic loss of GPS over an extended period of time (e.g., more than one month) and affecting a large geographic area(e.g., nationwide, continental, global), overall impact is more difficult to ascertain. Because of the diverse and highly distributed implementations of GPS-based solutions across the industry, any impact likely would be experienced in the form of a gradual degradation of network performance, with little potential for cascading network failures. The NSTAC also emphasizes that commercial communications networks do not operate in a vacuum, and service providersand network operators will take immediate corrective actions in response to any size event, particularly a large-scale catastrophic event with the potential to degrade the network. Even before all automatic means of backup are exhausted, companies will have already executed contingency plans and performed manual reconfigurations and network timing adjustments as required to maintain network operation.

Overall, industry members surveyed believe that their companies have taken measures to safeguard against those disruptions to the GPS signal that are likely to be encountered; however, to date, no industry or Government exercise has sought to replicate the impact of a long-term or permanent GPS outage simultaneously on all industries. The NSTAC recommends that the President direct the Department of Homeland Security and the Department of Defense to include various GPS outage scenarios in future planned disaster recovery exercises in coordination with the commercial communications industry.

Appendix A: Working Group Members, Government Personnel, and Other Participants

WORKING GROUP MEMBERS

Intelsat, Limited	Mr. Richard DalBello, Chair
The Boeing Company	Mr. Marc Johansen, Vice-Chair
Verizon Communications, Incorporated	Mr. James Bean, Vice-Chair
AT&T, Incorporated	Mr. Thomas Hughes
	Ms. Rosemary Leffler
Bank of America Corporation	Mr. Roger Callahan
The Boeing Company	Mr. William Patrick Reiner
	Mr. Robert Steele
Intelsat, Limited	Ms. Sallye Clark
	Mr. Sterling Winn
Lockheed Martin Corporation	Mr. Allen Dayton
National Cable & Telecommunications Association	Mr. Andy Scott
Qwest Communications International, Incorporated	Ms. Diana Gowen
	Mr. Thomas Snee
Science Applications International Corporation	Mr. Hank Kluepfel
Sprint Nextel Corporation	Mr. Lee Fitzsimmons
	Ms. Allison Growney
	Mr. John Stogoski
Raytheon Company	Mr. Bill Russ
Rockwell Collins, Incorporated	Mr. Ken Kato
Verizon Communications, Incorporated	Mr. Roger Higgins

OTHER PARTICIPANTS

Homeland Security Institute	Dr. Eric Sylwester
Volpe National Transportation Systems Center	Dr. James Carroll

GOVERNMENT PARTICIPANTS

Institute of Defense Analyses	Mr. Jim Doherty
Department of Homeland Security, National Communications System	Mr. Dale Barr
	Mr. Kelvin Coleman
Department of Homeland Security, United States Coast Guard	Captain Curtis Dubay
National Space-Based Positioning, Navigation, and Timing Coordination Office	Mr. Robert Crane
	Mr. Michael Shaw

Appendix B: Acronym List

BITS	Building Integrated Timing Supply
CDMA	Code Division Multiple Access
CO	Central Office
DAC	Digital Access and Cross-Connect
DHS	Department of Homeland Security
DOCSIS	Data Over Cable Service Interface Specification
DOD	Department of Defense
DTV	Digital Television
E911	Enhanced 911
FCC	Federal Communications Commission
GPS	Global Positioning System
HD	Hybrid Digital
HLR	Home Location Register
IBOC	In Band On Channel
IRIG	Inter-Range Instrumentation Group
LES	Land Earth Station
MSC	Mobile Switching Center
NIST	National Institute of Standards and Technology
NSTAC	National Security Telecommunications Advisory Committee
NTP	Network Time Protocol
PBX	Private Branch Exchange
PNT	Positioning, Navigation, and Timing
PRS	Primary Reference Source
PSAP	Public Safety Answering Point
PSTN	Public Switched Telephone Network
RF	Radio Frequency
SDH	Synchronous Digital Hierarchy
SME	Subject Matter Expert
SONET	Synchronous Optical NETwork
TDM	Time-Division Multiplexing
TSG	Timing Signal Generator
TT&C	Telemetry, Tracking, and Control
VLR	Visitor Location Register
VOD	Video On Demand

End Notes

[1] Within the commercial communications industry, the term "mitigation" often refers to **reactive** approaches or strategies applied after an event has occurred. Reflecting the industry responses collected for the study, this report uses the term more broadly to encompass both reactive approaches as well as **proactive**, or preventative, safeguarding strategies.

[2]To synchronize its clock, a GPS receiver requires signals from only one satellite in the constellation.

[3] Currently, there are 29 operational satellites and one experimental satellite in orbit.

[4]Pace, Scott, et. al. *The Global Positioning System: Assessing National Policies*, 1995.

[5]Available at the National Space-Based Positioning, Navigation, and Timing (PNT) Executive Committee Web site, *http://pnt.gov/policy*

[6]Source: National Space-Based PNT Executive Committee,*http://www.pnt.gov*

[7] *http://www.gps.gov*

[8]National Communications System, "Telecommunications Network Time Synchronization," NCS Technical Information Bulletin 99-4, April 1999.

[9] Under Phase II, the FCC requires wireless carriers, within six months of a valid request by a Public Safety Answering Point (PSAP), to begin providing the latitude and longitude of a caller, typically within 50 to 300 meters. The FCC requires carriers using GPS-enabled handsets to locate callers within 150 meters 95 percent of the time and within 50 meters about67 percent of the time. In September 2007, the FCC voted to require wireless operators to meet E911 requirements at a locallevel, specifically the jurisdictional areas of individual 911 PSAPs, expanding upon the previous statewide or multi-state level requirement. Carriers must meet location accuracy targets by September 11, 2012.

[10]The term "complete GPS loss" refers to the inability to receive any GPS signals at all locations. The term "localized GPS loss" refers to the inability to receive any GPS signals within a limited geographical area (e.g., campus, city, region). These terms are in contrast to the term "partial GPS loss" which refers to the inability to receive signals from some, but not all, GPS satellites. According to industry feedback, partial GPS loss has no appreciable impact on network operations and services.

[11] While the examination of network impacts related to specific GPS vulnerabilities is outside the scope of this study, the NSTAC considered information on a range of threats and vulnerabilities, including unintentional disruption (e.g., solar bursts and ionospheric interference, RF interference sources, human factors) and intentional disruption (e.g., shutdown, jamming, spoofing, and meaconing [a system of receiving radio beacon signals and rebroadcasting them on the same frequency to confuse navigation]). In reviewing these vulnerabilities, participants generally characterized the potential for any long-term complete GPS loss as unlikely. Participants also agreed that the further study on GPS vulnerabilities, particularly potential space weather impacts, is necessary to better characterize likelihood of impact to the commercial communications infrastructure (See Associated Press, *Solar Bursts Could Threaten GPS,* April 5, 2007).

[12] American National Standards Institute, *T1.101-1999: Synchronization Interface Standards for Digital Networks.*

[13]Lombardi, Michael, NIST Time and Frequency Division, *Legal and Technical Measurement* Requirements for Time and Frequency.

[14]As noted in the 2004 NSTAC study, a Stratum 2 timing source referenced by a Stratum 1 timing source will maintain accuracy for up to one month. As shown in **Table 2**, a Stratum 2 source with no reference performing at the **minimum** ANSI T1.101 accuracy requirement

(worst case) would result in an interval of about seven days between cycle slips. Similarly, at the minimum specified frequency accuracy for Stratum 3E clocks, the worst case interval between cycle slips is 104 minutes.

[15]Telcordia, Generic Requirements GR-253-CORE, *Synchronous Optical Network Transport Systems*, and GR-1244, *Clocks for the Synchronized Network,* among others.

[16] *NSTAC Satellite Task Force Report*, March 2004.

[17] Reception of satellite signals from four satellites is needed to establish accurate position.

[18] National Institute of Standards and Technology, Time and Frequency Division,*http://tf.nist.gov/timefreq/stations/wwvb.htm*

[19]Feedback from cable network operators also notes that the cable industry is deploying asynchronous links such as Gigabit Ethernet and will rely less on synchronous networks such as SONET in the future.

[20]In an oscillator, the "flywheel effect" refers to the continuation of oscillations after removal of the control stimulus.

[21]The Network Reliability Interoperability Council, an FCC advisory committee comprised of commercial communications companies as well as public sector stakeholders has developed generic network reliability and security best practices, including practices regarding GPS location accuracy for E911 service.*http://www.nric.org*.

INDEX

A

B

C

L

M

N

O

P

U

V

W

Y